Biology of Normal Proliferating Cells in vitro
Relevance for in vivo Aging

Interdisciplinary Topics in Gerontology

Vol. 23

Series Editor
H.P. von Hahn, Basel

KARGER

Basel · München · Paris · London · New York · New Delhi · Singapore · Tokyo · Sydney

Biology of Normal Proliferating Cells in vitro Relevance for in vivo Aging

Alvaro Macieira-Coelho
National Institute of Health and Medical Research (INSERM),
Paris, France

67 figures and 4 tables, 1988

KARGER

Basel · München · Paris · London · New York · New Delhi · Singapore · Tokyo · Sydney

Interdisciplinary Topics in Gerontology

Dr. A. Macieira-Coelho
Faculty of Health Sciences
Dept. of Cell Biology
S-581-85 *Linköping*

Library of Congress Cataloging-in-Publication Data
Macieira-Coelho, Alvaro, 1932
Biology of normal proliferating cells in vitro: relevance for in vivo aging/Alvaro Macieira-Coelho.
(Interdisciplinary topics in gerontology; vol. 23)
Bibliography: p.
Includes index.
ISBN 3-8055-4660-2
1. Cells-Aging. 2. Cell proliferation. 3. Cell culture.
I. Title. II. Series.
[DNLM: 1. Aging. 2. Cell Division. 3. Cell Survival. 4. Cells, Cultured. W1 IN679 v. 23/QH 608 M152b]
HQ1060.I53 vol. 23
[QH608]
362.6 s-dc19
[574.87'6]
DNLM/DLC
for Library of Congress 87-29710

Drug Dosage

The authors and the publisher have exerted every effort to ensure that drug selection and dosage set forth in this text are in accord with current recommendations and practice at the time of publication. However, in view of ongoing research, changes in government regulations, and the constant flow of information relating to drug therapy and drug reactions, the reader is urged to check the package insert for each drug for any change in indications and dosage and for added warnings and precautions. This is particularly important when the recommended agent is a new and/or infrequently employed drug.

Printed in Switzerland by Werner Druck AG, Basel
ISBN 3-8055-4660-2

Contents

Acknowledgements

We are indebted to Mr. Gonçalo Macieira-Coelho for organizing the data base of the bibliography of cell culture in aging research, and to Mrs. Brigitta Lindström for typing the manuscript.

I. Introduction

Rationale for the Use of Cell Culture in Gerontology Research

Although the use of tissue culture techniques to study aging started in the early 1920s, its generalization occurred only during the last 20 years. This did not take place without controversy, since many investigators did not realize that cells maintained in vitro can reveal many of the properties they display in the organism. This lack of knowledge led to the claim that the causes of aging in an animal cannot be approached by the study of cells in an artificial environment.

This view was surprising since there is an almost endless list of contributions to the understanding of human pathological conditions, from works involving the cultivation of cells, in particular of fibroblasts. Thanks to the facility with which one can grow this type of cells, it has been possible for instance to identify cytogenetic [Tjio and Nichols, 1985] as well as enzymatic [Davidson et al., 1963; Cleaver, 1969] defects of inherited diseases, to detect all sorts of receptors like those for hormones [Kaufman et al., 1975] and high and low density lipoproteins [Goldstein and Brown, 1974] and to establish their disturbance in pathological situations. Another illustration of the reproducibility in vitro of metabolic changes related to specific pathological conditions, is the finding that cultured fibroblasts from Progeria patients display an increased excretion of hyaluronic acid [Brown et al., 1985]; indeed, an increased urinary excretion of hyaluronic acid has been detected in this disease. These are just a few examples of the fundamental contributions made to human pathology from the cultivation of fibroblasts. In fact more than 100 pathological entities have been studied and further identified by the use of cultivated fibroblasts from the respective donors.

Evidence that somatic cells express in vitro many of the pathological aspects of the donor, comes also from the field of cancerology. This will be developed later herein, since it is one of the windfalls from the use of tissue culture in gerontology. In fact, it is now well established that somatic cells from some donors with cancer or at high risk of cancer present deviations from the pattern of physiologic aging at the cellular level [Azzarone et al., 1984a; Schor et al., 1985].

Hence, there is no reason to believe that the field of aging should not benefit from tissue culture as human physiopathology already has.

The first investigators to use cell culture in cytogerontology were probably Carrel and Ebeling [1921]. They compared the effect of serum from young and old donors on cultivated cells and found that the serum from young chickens supported better the growth of homologous fibroblasts than did that of old animals. There is a general consensus in tissue culture laboratories that the serum of newborn animals has a higher growth-promoting activity as compared to serum from older individuals. However, a recent report [Kondo and Nomaguchi, 1985] described results opposite to those of Carrel and Ebeling. This is quite surprising and it would be interesting if these data could be checked.

It was also Ebeling in 1921 who described a slower migration of cells from explants originating from older chickens as compared to those of younger animals. This parameter was later analysed again by other investigators [Soukupova and Holeckova, 1964; Waters and Walford, 1970; Bierman, 1978] who found that the latent period between explantation and cell migration increased in tissues from older animals.

These works described the decline with aging, of a fundamental aspect of cell behavior, i.e. cell movement. Motion and function are interdependent in a cell: events like cell migration, phagocytosis, pinocytosis, exocytosis, ligand-receptor interactions, transport and cell-to-cell communication, are examples of movement-dependent functions. These displacements at the cell periphery are transmitted within the cell through the lability of the components of the cytoskeleton, a structural complex which integrates space and function, influences the movement of organelles and molecules, and as we shall later see, can also be determinant for gene expression.

The loss of the capacity to migrate has been recently analysed and confirmed by other methods [Muggleton-Harris et al., 1982]. Monolayers of cultivated human fibroblasts from neonatal, adult and aged donors were 'wounded' and the migration of cells from the edge of the wound, was followed. It was found that the latent period prior to initiation of migration was correlated with the donor's age and with the in vitro population doubling level of the different cell populations. This work shows that in regard to this parameter, aging of cells in culture, in terms of number of doublings completed, has the same effect as aging in vivo.

Cell division is another movement-dependent event that has been extensively studied in experimental gerontology with cell culture techniques. An intimate relationship between movement and division has been described as early as 1919 by Loeb and Fleischer. They observed in fibroblast cultures that mitoses were more frequent where movement was more active and suggested that movement of cells were likely to lead to proliferation. This relationship between cell movement and cell division has been repeatedly found under different conditions although no explanation to the phenomenon was available. A mechanism has been recently proposed that establishes the link between cell motion and the initiation of the division cycle [Macieira-Coelho, 1983], which will be dealt with again later on, because of its relevance in the understanding of cellular aging. It has a bearing with the problem of the movement-dependent regulation of gene expression as will be described in another chapter.

The use of tissue culture in experimental gerontology was greatly stimulated by the work of Hayflick and Moorhead [1961] which dealt with the potential number of divisions of serially cultivated human fibroblasts. The problem of the growth potential of cells is crucial for aging, it concerns phenomena such as the regeneration of tissues, wound healing, the immune response and stem cell renewal.

It was known from the work of Swim and Parker [1957] that fibroblasts from certain species (e.g. human) stop growing after a period of rapid proliferation. Hayflick and Moorhead suggested that human fiblasts are endowed with a fixed number of doublings and that the exhaustion of their doubling potential is an expression of senescence at a cellular level (fig. 1). The evidence in favor of this concept is reviewed in the next chapter. Although it is known that hormones [Macieira-Coelho, 1966; Macieira-Coelho and Loria, 1974; Cristofalo and Kabakjian, 1975] and changes in the nutrient medium [Massie et al., 1974; Mellgren, 1975; Cooper and Goldstein, 1977; Schneider et al., 1978; Poiley et al., 1978; Ohno, 1979; Ryan, 1979a; Delinassios 1983; Okada et al., 1986; Mukherjee and Weinstein, 1986] can increase the number of cell generations, nevertheless according to the experience of hundred of laboratories, the life span of these cells is indeed limited. Physical agents such as fluores-

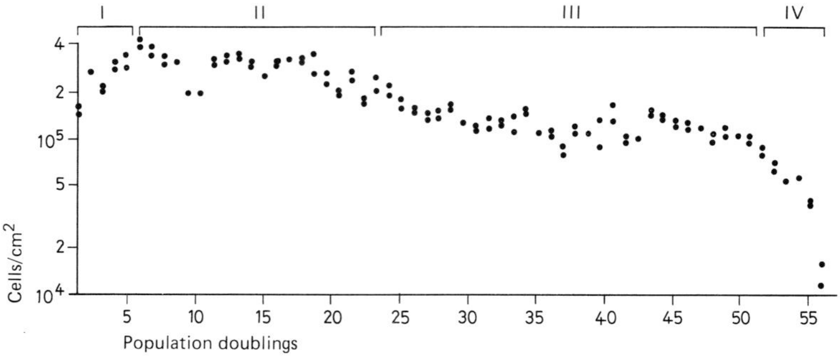

Fig. 1. Maximal cell densities recorded before each subcultivation during the entire life span of a human embryonic lung fibroblastic line (ICIG-7). The cultures were carried with a 1:2 split ratio. Up to the 39th doubling, cells were subcultivated at 3 to 4-day intervals. Thereafter and up to the 50th doubling, the population doubling time increased and the cells were subcultivated at 5 to 6-day intervals. After the 50th doubling it took 7 days to reach confluency. The last cell count was done 15 days after the previous subcultivation. The roman numerals indicate the approximate limits of the different phases which characterize the life span of the cell population [Macieira-Coelho and Taboury, 1982].

cent light and ionizing radiation can also extend the life span of cultivated cells [Litwin, 1972; Macieira-Coelho et al., 1977; Parshad and Sanford, 1977].

In summary, in vitro techniques raised the problem of circulating growth and inhibitory factors whose levels may vary in body fluids with aging of the organism, and provided a tool to reveal intrinsic cellular modifications which become apparent as a function of time and modify the behavior of cells. As will be shown through this monograph, in vitro cultivated cells express these changes of behavior.

These findings improved our knowledge in fields which extended well beyond the limits of gerontology. They had the advantage inter alia, of approaching human aging without endangering the individual and of making available a procedure that does not raise any ethical problems. For the moment tissue culture has mainly increased our knowledge concerning aging of dividing cell compartments. The opening, though, of new vistas in human physiopathology, has been immense.

Most of the work done so far, used fibroblast-like cells which seem to derive from the mesenchyme. But other cell systems have also been developed and will be mentioned herein accordingly.

The studies performed with cultivated lymphocytes will not be mentioned since they implicate the immune response of the organism, a subject that would deserve in itself a monograph.

Only experimental approaches will be analysed in this volume. Theories will be mentioned but not be described since they have been reported in detail in several instances in the literature.

II. Relationship between the Behavior of Cells in Culture and the Physiopathology of the Organism

1. Development, Aging, Disease and the Potential to Divide in vitro

The idea that somatic cells may have a limited division potential and that it may have implications for aging of the organism, was the subject of theoretical considerations made by Weissman [1891] in the last century. Hayflick's [1961] work was the first experimental approach to the problem. His concept was reinforced by his own finding that fibroblasts from adult human donors have a smaller division potential than cells of embryonic origin [Hayflick, 1965]. This work did not establish definitely a relationship between aging and the loss of the capacity to divide. It proved though that there is a difference between the growth potential of embryonic cells and that of postnatal cells, in other words, that during development there is a loss in the doubling potential of cells.

This loss does not occur at the same rate in all tissues [Schneider et al., 1977]. The growth potential of skin fibroblasts for instance declines at a faster rate than that of lung fibroblasts. This is illustrated in figure 2 which shows the number of doublings of skin and lung fibroblasts from the same embryo, cultivated in vitro. It could be due either to different turnover rates or to different evolutions of the fibroblast stem cell compartment in the respective tissues.

Within the same tissue, differences in the growth potential of fibroblasts have also been described. In the skin, for instance, fibroblasts from the papillary layer have an increased survival in vitro when compared to reticular fibroblasts [Harper and Grove, 1979; Azzarone and Macieira-Coelho, 1982; Schafer et al., 1985]. These differences in the capacity to divide are accompanied by distinct biochemical features. In fact, the pattern of the synthesis of surface proteoglycans is not the same in these two types of fibroblasts [Beyth and Culp, 1985]. It is not known, however, if these two cell types represent different stages of the evolution of the same stem cell compartment or separate compartments. In vivo and in vitro

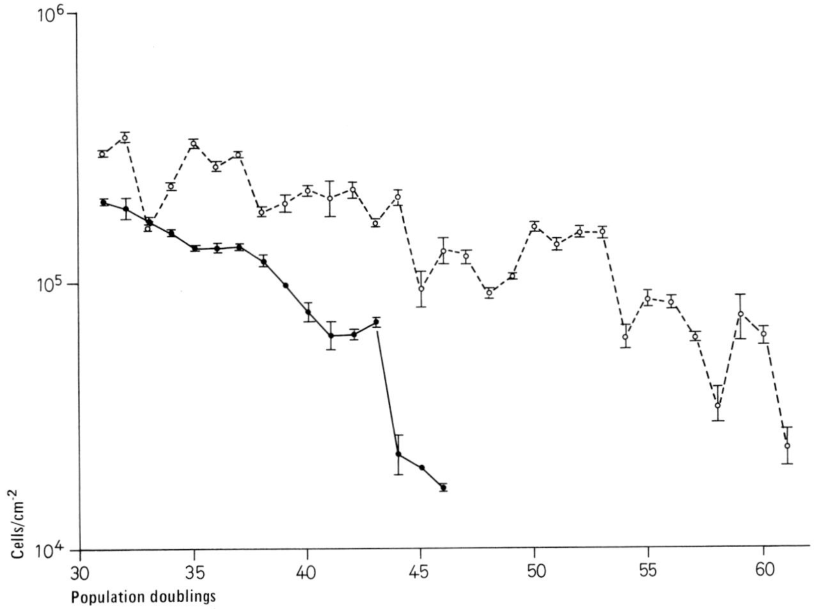

Fig. 2. Maximal cell densities before each subcultivation of fibroblast populations derived from the lung (----) and skin (—) of the same human embryo. Cell counts were started after the 30th doubling.

aging of these two types of fibroblasts resulted in different alterations of the metabolism of proteoglycans [Kent et al., 1986].

Additional evidence that indeed serial divisions of cells in vitro cause changes identical to those taking place during development, came from the finding that the kinetics of proliferation occurring in embryonic cells after serial proliferation in vitro, are identical to those found in postnatal cells [Macieira-Coelho and Pontén, 1969].

Results buttressing the relationship between the loss of the division potential and aging, were obtained by Martin et al. [1970]. They found a decline of 0.2 doublings per year among fibroblast populations originated from donors of different ages. These results have been challenged by Goldstein et al. [1978] who could not find a significant negative correlation between donor age and the doubling potential in vitro of fibroblasts originated from normal donors. A significant negative correlation though, was present with fibroblasts from diabetic patients or from donors genetically predisposed to diabetes. The authors concluded that the physiopathologic status of the tissue's donor is a more precise determinant of fibroblasts replicative life span than chronological age. It is possible that the negative correlation was not significant for the normal donors in Goldstein's experiments, because the number of donors was not significantly large. It is also possible that the skin biopsies were not well standardised so that fibroblasts from different layers which are known to have different life spans (see above) were propagated in vitro. Another source of error comes from the finding that the division potential of fibroblasts from some cancer patients does not correlate with the age of the donor [Azzarone et al., 1984]. If among the donors, some had cancer, the number of doublings of their fibroblasts would disturb the correlation.

Abnormalities in proliferation of fibroblasts from diabetic patients were also reported by Rowe et al. [1977]. Within their nondiabetic controls, they found an inverse correlation between donor's age and the doubling rate and maximal cell densities of the respective fibroblasts. Fibroblasts from both juvenile- and adult-onset diabetes had a reduction of those two growth parameters. In addition, hydrocortisone caused a reduction of protein and collagen synthesis in cells from juvenile-onset but not from adult-onset diabetes.

Other investigators did not find any decrease in the proliferative potential of cells from diabetic patients [Vracko and McFarland, 1980; Germinario et al., 1986]. The variations, however, of the cumulative replicative life spans of cells from diabetic donors, was greater than among control cells [Vracko and McFarland, 1980]. It should be pointed out though that there are different types of diabetes and that the patients pooled might not be comparable.

So one may ask whether the division potential of fibroblasts in vitro concerns only changes in development and disease or if it also concerns changes due to aging. In any case, these three processes are interrelated, and if it is well proven that tissue culture can be used to study development and disease, this in itself justifies its use in aging research. We believe that the possibility of studying the interactions of these three phenomena is one of the advantages of the tissue culture model.

Other authors have also found a correlation between the long-term doubling potential of human cells in vitro and the donor's age. This was so, not only for mesenchymal cells of different tissues [Le Guilly et al.,

1973; Schneider and Mitsui, 1976; Vracko et al., 1983; Mets et al., 1983; Azzarone et al., 1984; Lipschitz and Udupa, 1984; Bartold et al., 1986] but also for lens [Tassin et al., 1979], arterial smooth muscle [Bierman, 1978; Volicer et al., 1983], keratinocytes [Rheinwald, 1979], chondrocytes [Evans and Georgescu, 1983] and a population of T lymphocytes [Walford et al., 1981]. An interesting test of Hayflick's hypothesis was made by Ryan et al. [1981] who cultivated skin fibroblasts from pairs of monozygotic twins. The results showed a similar proliferation and replicative life span within each twin pair. Among the twin pairs, however, the cells differed in these respects.

A decline in the cell growth potential with age was also found for fibroblasts from tortoise [Goldstein, 1974] and hamster [Bruce et al., 1986], for smooth muscle cells from two murine species [Martin et al., 1983] and for chondrocytes from dog and rabbit [Adolphe et al., 1983; Evans and Georgescu, 1983].

This is indeed an impressive list of works buttressing Hayflick's suggestion.

External factors such as sun exposure, can affect the division potential of skin fibroblasts. It was found [Gilchrest, 1980] that fibroblasts derived from exposed skin had a shorter division potential than those from the nonexposed one of the same donor. This shortening of the doubling potential by exposure to sunlight became more pronounced with donor age and the severity of clinical aging changes. In other words, aging and sun exposure have a potentiating effect on the shortening of the fibroblast division potential. A decreased life span in vitro of fibroblasts derived from skin exposed to photochemotherapy in vivo has also been reported [Jacobson-Kram et al., 1982].

Other parameters besides the long-term division potential have also been tested. Thus, human epidermal keratinocytes from newborn donors proliferate more rapidly and attain a higher concentration at confluence than those from aged donors [Liu et al., 1982]. Also during development and aging, there is an increase in the concentration of PDGF, required for optimal growth in vitro [Slayback et al., 1977; Plisko and Gilchrest, 1983; Nilsson et al., 1983; Phillips et al., 1984].

Further support for the relationship between aging and division potential came from studies with cells originated from patients with Werner's syndrome of premature aging. These cells have a decreased doubling potential when compared with those of age-matched normal donors [Salk et al., 1981].

Another disease with features of premature aging, where fibroblasts were found to senesce earlier in vitro than age-matched controls, is cystic fibrosis [Shapiro et al., 1979].

A question that has been raised is if the limited life span of the cells is due to the number of doublings or to the calendar time spent in vitro. This has been settled with an elegant experiment performed by Dell'Orco et al. [1973] who could maintain human fibroblasts for long periods in a non-dividing state in a medium supplemented with 0.5% serum; when the cells were again serially replicated they went through the same number of doublings as the control group which had not been maintained in resting phase and had thus spent a shorter calendar time in vitro. Another type of experiment supported these findings [Brunk et al., 1973; Harley and Goldstein, 1978]. Colonies of cells were maintained without transfer so that cells in the center were in resting phase during all the experimental time while proliferation occurred only at the periphery. It was found that the peripheral cells exhausted their division potential contrary to the cells at the center which could still divide.

Obviously, it would be naive to think that in old age our somatic cells have lost their capacity to divide. This has been a misinterpretation due in part to the approach of some investigators which have considered senescent cells, only those that are in the terminal stage of their life span. One of the contributions of the cell culture model is to have shown [Macieira-Coelho and Taboury, 1982; Macieira-Coelho and Azzarone, 1982] that although some terminal cells are produced from the mass population at each doubling, long before the cell population reaches the terminal stage, small changes occur in the initiation of the division cycle and in the transit through the cycle. It is these progressive modifications that can explain such events like the delayed wound healing [Holm-Pedersen, 1974] and the disturbed modulation of intestinal cell production [Holt et al., 1987] observed during aging in vivo; they are certainly the expression of the modifications created in the cell through division, which are apparent long before a cell reaches the terminal postmitotic state. In other words, events taking place during the division cycle, which modify the cells and which will be detailed later, create a permanent drift that leads to heterogeneity and to changes in the response of the cells to growth stimuli.

Moreover, it should not be forgotten that the capacity to divide is just one cell function among others and that if it is impaired, it must mean that other functions are also impaired. Concerning fibroblasts for instance which are responsible for the synthesis of supporting structures like collagen, proteoglycans, structural glycoproteins and elastine [Robert and Robert, 1973] and in general for the creation of a microenvironment through the synthesis of diffusable substances, a shift in cell functions will certainly disturb the mutual relationships with other cell systems and the metabolism of the latter as well. These type of events are certainly important in the physiology of aging.

In pathological conditions, however, the final stage of the cells' growth potential could constitute an important aspect of the pathogenesis of the disease at the local tissue level. This may be relevant for atherosclerosis were smooth muscle cells from the fibrous plaques of arteries have a decreased growth response to whole-blood serum, analogous to that observed among senescent cells in culture [Ross, 1986]. Another interesting aspect in regard to the pathogenesis of atherosclerosis has been raised by Martin et al. [1974]. They suggested that some disorderly proliferations leading to local hyperplasia during aging and playing a role in different conditions including atherosclerosis, are due to a dysequilibrium between cellular systems, the decline of the growth potential of one leading to the overexpression of another. Pertinent to Martin's suggestion is the finding of Mets and Verdonk [1981] who found that the loss of human bone marrow fibroblast-like progenitor cells during aging is compensated by an increase in the number of epithelial-like cells.

Other markers have also been used to detect developmental and age-dependent changes in mesenchymal cells using tissue culture techniques. Several markers were found to concern development rather than aging. Thus, a decreased activity of the hexose monophosphate shunt has been found in postnatal fibroblasts as compared to embryonic ones [Condon et al., 1971], which could be responsible for the decreased proliferative potential. Moreover, Clemmons [1983] has reported that human fibroblasts lose the capacity to secrete autocrinal growth factors during development and become more dependent on the external supply of such factors.

Other ontogenic related changes in the response to growth factors, have been reported. Addition of hydrocortisone to serum-supplemented medium stimulates the growth rate and final cell density of skin fibroblasts from young adults but not from fetal subjects [Rowe et al., 1977]. Also while in adult dermal fibroblasts dexamethasone acts synergistically with somatomedine C, no such interaction exists in fetal fibroblasts [Conover et al., 1986]. In the former cells, the activity of 10% calf serum can be fully duplicated by the combination of four growth factors (somatomedine C, EGF, PDGF and dexamethasone); in fetal cells, however, dexamethasone has little effect and the combination of growth factors replaces only 50% of the activity of 10% calf serum. Furthermore, these authors found that the fetal cells are more responsive than postnatal fibroblasts to the mitogenic effect of serum.

At the membrane level, developmental rather than aging-related changes can also be detected through the cultivation of cells. They concern differences in agglutinability [Cassiman et al., 1977] and the masking of surface antigens in postnatal cells as compared to embryonic ones [Azzarone et al., 1984b]. They could be due to architectural modifications of the cell membrane. Another developmental change that could be due to modifications of the cell membrane is the resistance to coxsackie B4 infection caused by a block in virus penetration. Skin fibroblasts from young fetuses are susceptible to coxsackie B4 infection, whereas those from older fetuses, children and adults are not [Hornberger and Plotkin, 1983].

Changes in ^{60}Co and X-ray sensitivity [Macieira-Coelho et al., 1977, 1978; Beverstock and Simons, 1982] also occur during development.

Other markers that are expressed in vitro are clearly age-related. The list of events (table I), besides those already mentioned above, that were found to accompany in vitro as well as in vivo aging is a long one; they give further support to the validity of the tissue culture model in gerontology research.

It should not be expected, though, that all changes taking place during in vitro aging are identical to those occurring in vivo. The cultivation of cells must amplify certain events since it imposes on the cell population a continuous proliferation which does not take place in the organism, and must correspond to a stress situation. Adult smooth muscle cells, for instance, when put into culture, undergo a modulation similar to that occurring in vivo in tissues proliferating in response to injury [Ross et al., 1986]. In addition to the differences between in vivo and in vitro aging already mentioned above, others have been reported. Thus, the glucocorticoid amplification of insulin-stimulated DNA synthesis, which is depressed with increased donor age, does not decline with age in vitro [Germinario and McQuillan, 1985]. Changes in the requirement for nutrients also do not always evolve in the same way: newborn cells grow in both serum and platelet-poor plasma; adult cells grow only in whole blood serum; newborn cells, however, when they age in vitro, do not acquire adult-type requirements [Wharton, 1984]. Differences in aging in vitro and in vivo have also been reported in the metabolism of sulfated

Table I. Events occurring with in vitro and in vivo cell aging

Loss of the division potential (see text for references)
Chromosome alterations [Saksela and Moorhead, 1963]
Loss of the capacity to migrate [Soukopova and Holeckova, 1964; Waters and Walford, 1970; Muggleton-Harris et al., 1982]
Prolongation of the G2 period [Macieira-Coelho, 1967a; Gelfant and Smith 1972; Kubbies et al. 1985]
Loss of the hybridization signal for the probes of some genes [Johnson and Strehler, 1972; Icard-Lipkalns et al., 1986]
Decreased mitogenic response to growth factors [Slayback et al., 1977; Plisko and Gilchrest, 1983; Phillips et al., 1984]
Increase in cell size [Simons and van den Broek, 1970; Lee et al., 1978]
Colony size distribution [Smith et al., 1978; Steinhardt, 1985]
Changes in collagen synthesis [Basler et al., 1979]
Increase in the super oxide dismutase specific activity [Duncan et al., 1979]
Decline in induced sister chromatid exchanges [Schneider and Gilman, 1979; Reimer and Singh, 1983]
Changes in lectin-binding sites [Aizawa et al., 1980b; Aubery et al., 1980]
Decreased cell spreading [Kelley et al., 1980; Peterson et al., 1986]
Decreased cell-fusion potential [Röhme, 1981]
Replicative potential of cell hybrids derived from young and old cell populations [Muggleton-Harris and Aroian, 1982]
Decreased binding, uptake and degradation of low density lipoproteins [Lee et al., 1982]
Chromatin changes at the level of the 30- and 10-nm fibers [Puvion-Dutilleul and Macieira-Coelho, 1983]
Changes in low density lipoproteins metabolism [Bierman et al., 1979; Bierman and Schwartz, 1984; Behrman and Wynn, 1984]
Decreased contractility as measured by fibrin clot retraction [Niewiarowski and Goldstein, 1973; Azzarone and Macieira-Coelho, 1984]
Presence of extrachromosomal circular DNA [Puvion-Dutilleul and Macieira-Coelho, 1983; Kunisada et al., 1985; Icard-Liepkalns et al., 1986]
Decrease in the relative amount of vimentin in lens cells [Ramaekers et al., 1984]
Decreased ATP content in the presence of metabolic poisons [Muggleton-Harris and Defuria, 1985]
Increase in heparan sulfate and a relative decrease in dermatan and chondroitin sulfate content [Schachtschabel and Wever, 1978; Bartold et al., 1986]
$CaCl_2$ dependency of contact inhibition of growth [Praeger and Gilchrest, 1986]
Increased labeling of proteins derived from putrescine via the polyamine catabolic pathway [Chen and Chang, 1986]
Alteration of the thyroid hormone induction of (Na+K)ATPase [Guernsey et al., 1986]

proteoglycans by chicken chondrocytes [Pacifici et al., 1981]. Another difference concerns the effect of interferon. The concentration of the latter, required to induce the antiviral genes in fibroblasts from older donors, is 2 to 3-fold lower than that required to induce the genes in cells from younger donors [Tan et al., 1975]. With aging in vitro, however, there is no change in the regulation of antiviral genes expression.

We do not want to finish this section without emphasizing that it is obvious from some of the examples given above that several factors other than cell divisions can decrease in vivo the division potential of cells. It is possible that many of the changes observed in cells that partially lost their potential to divide are the same regardless of the events that caused that loss.

2. Aging, Cancer and the Life Span of Cells in vitro

One pitfall when making interspecies comparisons in regard to growth potential arises from the tendency of cultivated fibroblasts from some species to overcome the growth decline and acquire the capacity to replicate indefinitely. Indeed, the investigator comparing life spans of fibroblast-like cells may have the feeling of going 'through a looking glass' and penetrating a nonsense world. Chicken embryonic fibroblasts, for instance, invariably die after a certain number of doublings [Lima and Macieira-Coelho, 1972]. On the other hand, mouse cells behave the opposite way and invariably become spontaneously permanent cell lines, i.e. they immortalize [Todaro and Green, 1963]. Cells from other rodents can spontaneously develop permanent lines with a variable frequency; rat fibroblasts also yield easily permanent cell lines and in regard to hamster fibroblasts the probability of obtaining an immortalized cell line from Syrian hamster is higher than that from rabbit but less than for Chinese hamster or rat [Terzi and Hawkins, 1975]. These features have raised the problem whether the different growth potentials can be attributed to nature or to nurture. Some investigators are inclined to accept that these differences are due to the fact that the nutrient media available are adequate for those cells that can divide indefinitely but do not provide all the required nutrients for those cells which die after serial doublings [Moore and McLimans, 1968].

Several arguments though suggest that the different behaviors of these cell populations are due to their intrinsic properties.

A strong argument in favor of the idea that indeed the nutrient media are adequate for the maintenance of the cells with a limited life span was provided by the results of Dell'Orco [1974] mentioned above, showing that one can maintain human fibroblasts for an indefinite period in resting phase without losing their division potential.

The other argument in favor of nature rather than nurture is the fact that the different probabilities of limited or unlimited doubling potentials have a correlation with the response of the respective cell populations to the action of carcinogens, whether viral, chemical or physical, and of oncogenes.

Immortalization is one of the steps in the complex phenomenon called transformation which is the counterpart in vitro of malignancy in vivo [Hayflick, 1967]. Transformation consists in the progressive acquisition of growth autonomy such that a cell population becomes less dependent upon the different mechanisms which restrain the growth of normal cells in vitro. During transformation a cell population goes through a serie of steps whose succession is unpredictable and which can stop short of full autonomy, i.e. full transformation. The extent of transformation of the different type of fibroblasts when they are treated with carcinogens has a striking bearing with the respective tendency of having spontaneously a limited or unlimited division potential.

Thus, chicken fibroblasts infected with Rous sarcoma virus (RSV), although transformed in the sense that they lose contact inhibition of growth, have a shorter life span than identical populations of noninfected cells [Pontén, 1970] and although it is possible to obtain permanent cell lines from the infected cells [Géder et al., 1973; Dinowitz, 1977], the probability is low. Furthermore, the sarcomas induced by these transformed cells are not transplantable and seem to propagate by infection of new cells which die after a few divisions [Bergs and Groupe, 1963; Pontén, 1964]. Mouse sarcomas obtained by the in vitro infection of cells with oncogenic viruses, however, are transplantable and present the type of clonal growth where all cells in the tumor are derived from the first transformed cells.

Human cells seem to have an intermediate behavior in regard to the response to oncogenic viruses when compared with chicken and mouse. When infected with RSV they lose contact inhibition of growth but do not immortalize [Stenkvist, 1966], if infected by SV_{40} virus they also lose contact inhibition of growth and can yield with a low probability immortal cell populations [Pereira-Smith and Smith, 1981].

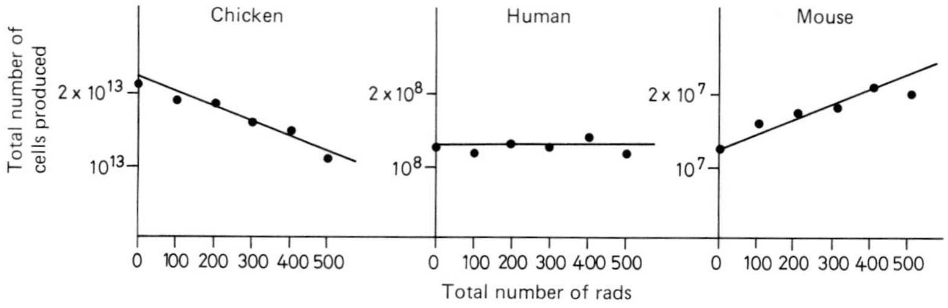

Fig. 3. Total amount of cells produced in fibroblast control cultures from the indicated species and in identical cultures irradiated with the different doses of ionizing radiation. In mouse fibroblasts the total number of cells produced was calculated up to passage 20, that is, at the time when the saturation densities became stabilised [Macieira-Coelho et al., 1976].

The relative position of these cell types in regard to the response to oncogenes is identical. Thus, oncogenes can accelerate the immortalization of rat cells [Land et al., 1983], a phenomenon which anyway occurs spontaneously with these cells, they do not immortalize human fibroblast [Sager et al., 1983] and they transform without immortalization chicken cells [Stehelin et al., 1976].

The same type of response is found with ionizing radiation. Indeed, radiation accelerates the immortalization of mouse cells [Macieira-Coelho et al., 1976], it may have no effect, prolong the life span [Macieira-Coelho et al., 1977] or, with a low probability, immortalize human cells [Namba et al., 1985], and it shortens the division potential of chicken fibroblasts (fig. 3) [Macieira-Coelho et al., 1976].

The same response was observed with chemical carcinogens. It is easy to obtain full transformation of mouse or rat fibroblasts with these substances but it requires a special schedule to obtain it with human fibroblasts [Milo and Casto, 1986]. In the latter also the process stops short of full transformation.

One can thus produce a scale (fig. 4) with at one end cells like chicken fibroblasts which very rarely give origin to permanent cell lines even in response to carcinogens, and at the other end fibroblasts which spontaneously produce immortal populations and where carcinogens and oncogenes are just accelerators of a latent potential. Finally, in between are fibroblasts where carcinogens can prolong the life span but more rarely give rise to immortal populations.

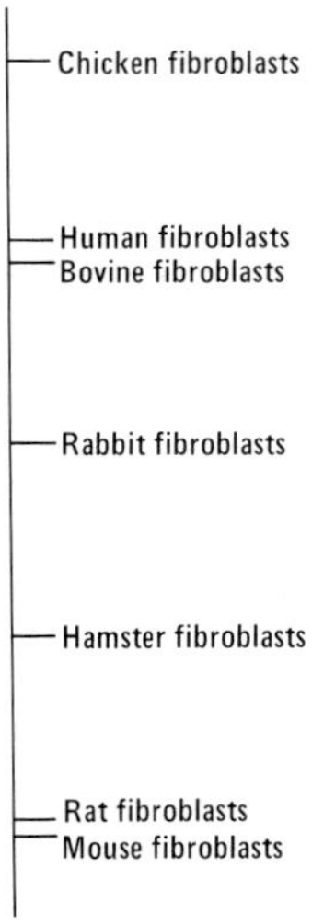

Fig. 4. Scale representing from top to bottom the increasing probability for fibroblasts from different species to become spontaneously permanent cell lines [Macieira-Coelho, 1980].

Pertinent to the problem dealt herein is the fact that it is easier to induce cancers in those species whose fibroblasts have a higher probability of escaping senescence [Pontén, 1971].

Another finding showing that aging and transformation at the cellular level represent two opposing forces is the fact that the division potential of fibroblasts from some cancer patients does not decline with the age of the donor, as is the case for the cells of normal donors (fig. 5) [Azzarone et al., 1984a]. The fibroblasts from the cancer patients also show in vitro other deviations from normalcy, suggesting that they are less influenced by the mechanisms that restrict the growth of normal cells in vitro [Azzarone et al., 1984a]. It has been suggested that this is due to the leftovers of an embryonic phenotype in these mesenchymal cells with the creation of a field effect that facilitates the transformation of epithelial cells [Schor et al., 1985]. A prolongation of the life span of fibroblasts from donors at high risk of cancer through variations in the culture conditions has also been reported [Diatloff-Zito and Macieira-Coelho, 1982].

Another analogy between interspecies differences in fibroblast behavior and differences between human fibroblasts from different human donors concerns the response of the latter to viral, chemical and

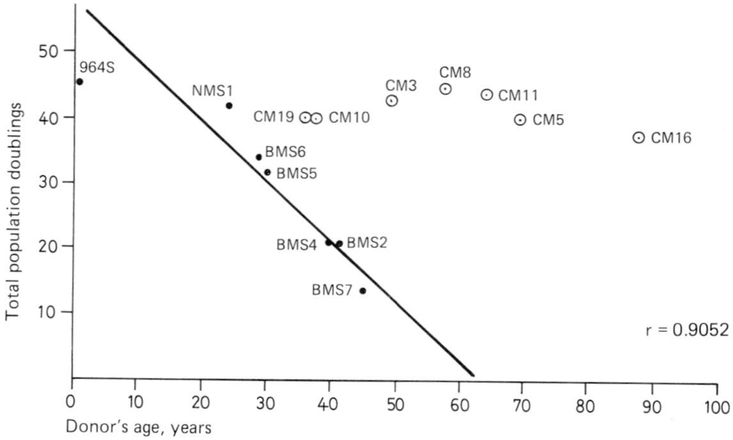

Fig. 5. Age of the donor versus average population doublings of the respective skin fibroblasts, derived from a human embryo (964 S) and from patients with benign tumors (●) or breast cancers (○). The straight line is the regression fitting the plot corresponding to the fibroblasts originating from donors with benign lesions. The correlation coefficient is indicated on the chart [Azzarone et al., 1984a].

physical carcinogens. Indeed, skin fibroblasts from human donors with cancer or at high risk of cancer seem more susceptible to transformation by carcinogens than those obtained from normal donors. Although immortalization of human cells is a rare event, the doubling potential of fibroblast from donors with cancer [Rhim et al., 1980; Azzarone et al., 1980] or at high risk of cancer [Diatloff and Macieira-Coelho, 1979] can be prolonged by chemical carcinogens and by radiation. The cells from the latter patients also seem to be more susceptible to viral transformation [Rhim et al., 1980].

These features which are fundamental for the physiopathology of cancer have been one of the windfalls from the use of the tissue culture model in cytogerontology. They show that the behavior of cells in vitro has a baring with the physiopathology of the organism and is not just a problem of finding the right nutrients. They argue strongly against theories claiming that the finite life span of cells in vitro is the result of culturing procedures. One of these theories, known under the name of commitment theory, proposes that cells committed to senescence are produced with a fixed probability at each division of the potentially immortal cells. The methods used to subcultivate the cells, would select for committed cells, the uncommitted ones being progressively lost [Holliday et al., 1977]. However, the extrapolation of cumulative growth curves during immortalization of mouse cells, suggest that the immortalized cells are not present in the primary culture [Todaro and Green, 1963]. Furthermore, experiments with caffeine showed that mouse cells during the process of immortalization become resistant to the drug and that the resistant cells are not present in the initial cell population [Icard and Macieira-Coelho, 1981]. The direct follow-up of the appearance of immortal rat cells with cinematography also suggested that the cells acquiring an unlimited division potential are not present in the primary culture [Matsumura et al., 1985].

Another question that is raised by this different behaviors of cells in vitro, that has implications for aging, is the fact that these species, where it is easier to induce tumors and whose fibroblasts are easily transformed in vitro, are also short lived. Have these characteristics of the somatic cells any implications for the shorter life spans? To the best of our knowledge a comparison between the frequency of spontaneous tumors in short- and long-lived mammals has not been done. Other mechanisms, however, could be involved in the curtailing of the life span, besides the tumor frequency.

Germane to this relationship is the correlation found between the species life span and the capacity of the respective cultured fibroblasts to activate a carcinogen into a mutagenic form and to metabolize the carcinogen into a water-soluble form via enzymes of the microsomal fraction [Schwartz and Moore, 1977]. Moreover, a good correlation was found [Schwartz, 1975; Moore and Schwartz, 1978; Pashko and Schwartz, 1982] between species' life span and the cytochrome P-448 content of the cultured fibroblasts, which is a major component of the microsomal metabolizing fraction. In other words, shorter-lived species have a higher content of cytochrome P-448. These results could be in favor of the possibility that the life span of some species is curtailed by a higher incidence of tumors. They could also mean that the same mechanisms involved in the activation of carcinogens and in detoxification are also implicated in senescence.

These findings also provide other hypotheses to explain how fibroblasts from some individuals could create a microenvironment favorable for the transformation of epithelial cells and open a whole new concept in the physiopathology of the neoplastic disease.

3. The Division Potential of Cells in the Organism

The problem if cells in vivo have a limited life span has been the subject of much debate. The experiments suggest that they do, but that it may not be relevant for aging in all tissues.

Daniel and Young [1971] have analyzed the division potential of serially transplanted mouse mammary glands and found that this cell system has indeed a finite doubling capacity.

Hemopoietic cells also have a limited division potential [Siminovitch et al., 1964; Cudkowicz et al., 1964] but the latter, it has been claimed, is not affected by aging [Harrison et al., 1978; Schofield et al., 1986].

A finite proliferative potential has also been reported for an antibody-forming cell clone after serial transplantation [Williamson and Askonas, 1972]. Thus, it is possible that in some circumstances the limited proliferative potential of the haemopoietic tissue also has implications for aging. Indeed, a decreased proliferative response during aging of the organism has been found with lymphocytes [Hori et al., 1973; Price and Makinodan, 1972] and is probably related with the delayed immune response well known to take place in old individuals. Maybe in the hemopoietic tissue, there is no growth decline of stem cells but only of differentiated cell types with aging.

Other tissues where a decline in proliferation has been reported in vivo during aging of the animal are rat mast cells [Norrby et al., 1981] mouse erythropoietic cells [Udupa and Lipschitz, 1984], ear and plantar epidermis, cells from cortical tubules of the kidney, ventral epithelium of the tongue, epithelium of the esophagus and epithelium of the colon [Cameron, 1972]. A decline in the proliferation of intestinal epithelial cells previously reported [Lesher et al., 1961] could not be confirmed [Holt et al., 1984]. It seems instead that in this tissue, it is the regulation of proliferation after a stimulus that is disturbed [Holt et al., 1987]. The difference in results seems to be due to the use of germ-free animals by the latter authors. Maybe in Lesher's work could the presence of intestinal infections constitute a continuous stimulus that triggered a decline of cell proliferation. In one instance (mouse prostate) the decline in mitotic activity with age seemed to be due to extracellular factors, e.g. hormones [Simnett and Morley, 1967].

In humans, a decline in epidermal cell renewal during aging of the skin has also been observed [Grove and Kligman, 1983].

Another question that has been raised is if there is any relationship between the division potential of somatic cells in vitro and the life span of the respective species. According to some investigators there is no correlation [Stanley et al., 1975]; others claimed to have found such a relationship [Röhme, 1981]. Unfortunately, in the former publication the cells from the different species did not originate from the same tissues and in the latter embryonic cells were compared with postnatal ones. Furthermore, in the case of embryonic cells the whole embryo was used, which gives rise to highly heterogeneous cell populations. Hence, as long as the same types of cell are not used for comparison, the question remains unsettled. Martin et al. [1983], however, compared the cloning efficiency of smooth muscle cells originated from two murine species with different life spans and found a twofold increased cloning efficiency in the longer-lived species.

III. Putative Mechanisms Responsible for the Different Growth Potential of Somatic Cells

1. Reorganization of the Genome

Several experiments suggest possible explanations for these different behaviors of cells in vitro. Some of them concern observations at the cytogenetic level during serial divisions of these cell populations. They suggest that the long-term doubling potential of cells is directly related with the potential for continuous chromosome rearrangements, in other words, with the 'plasticity' of the genome.

Human embryonic fibroblasts when cultivated in vitro go through continuous chromosomal rearrangements without any definite pattern becoming predominant [Chen and Ruddle, 1974; Harnden et al., 1976]. The data suggested that multiple clones arise continuously and compete between each other without any one overgrowing the others. These cells have a longer division potential than postnatal fibroblasts which go through more stable clonal-type chromosomal rearrangements during serial divisions in vitro [Harnden et al., 1976]. Moreover, fibroblasts from Werner's syndrome patients, which have a reduced doubling potential when compared with cells from age-matched normal donors, present chromosomal rearrangements (variegated translocation mosaicism) which become predominant and fixed during the cell population in vitro life span [Salk et al., 1981]. These findings at the cytogenetic level suggest that a 'rigidity' of the genome is associated with a shorter doubling potential as opposite to a higher 'plasticity' which would lead to a prolonged cell life span.

Another pertinent finding is the one showing that sister chromatid exchanges in human fibroblasts, which have used most of their doubling capacity, are less stimulated by drugs known to induce them [Schneider and Gilman, 1979]. These results suggest that the loss of the division potential is associated with a loss of the 'plasticity' of the genome. In lymphocytes from progeria and Werner's syndrome patients, however, mitomycin C-induced sister chromatid exchanges were not different from the normal controls [Darlington et al., 1981].

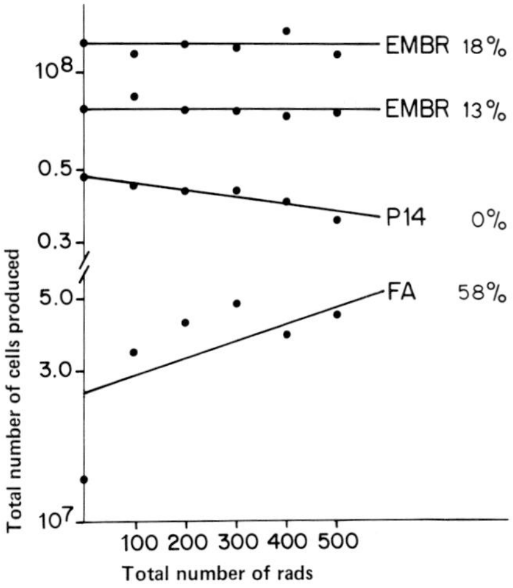

Fig. 6. Total amount of cells produced in control human fibroblast cultures and in identical cultures irradiated with the different doses of ionizing radiation. EMBR = Embryonic lung; P14 = adult lung fibroblasts; FA = cells from a Fanconi's anemia patient. The % indicates the frequency of chromosome rearrangements observed in the respective nonirradiated cells through their in vitro life span.

The study of the response of the different types of cells to ionizing radiation gave some clues to the intriguing relationship between response to carcinogens and spontaneous long-term growth potential, described above. Human embryonic fibroblasts when irradiated with low dose rate radiation at low population doubling levels (PDL) are more resistant and their life span can even be prolonged [Macieira-Coelho et al., 1977; Croute et al., 1986]. When irradiated late during their life span, however, their doubling potential is shortened. The same type of radiation shortens the division potential of early passage postnatal fibroblasts, and it seems that these different effects are related to the 'plasticity' or the 'rigidity' of the cell genome [Bourgeois et al., 1981]. In other words, the control embryonic cells went through more rearrangements than the postnatal cells in agreement with the findings of Harnden et al. [1976] reported above. On the other hand, the same type of radiation can prolong the life span of human postnatal fibroblasts from donors with genetic diseases with a high rate of chromosome recombinations, which are at high risk of cancer [Diatloff and Macieira-Coelho, 1979]. These findings are illustrated in figure 6 which shows the effect of different doses of ionising

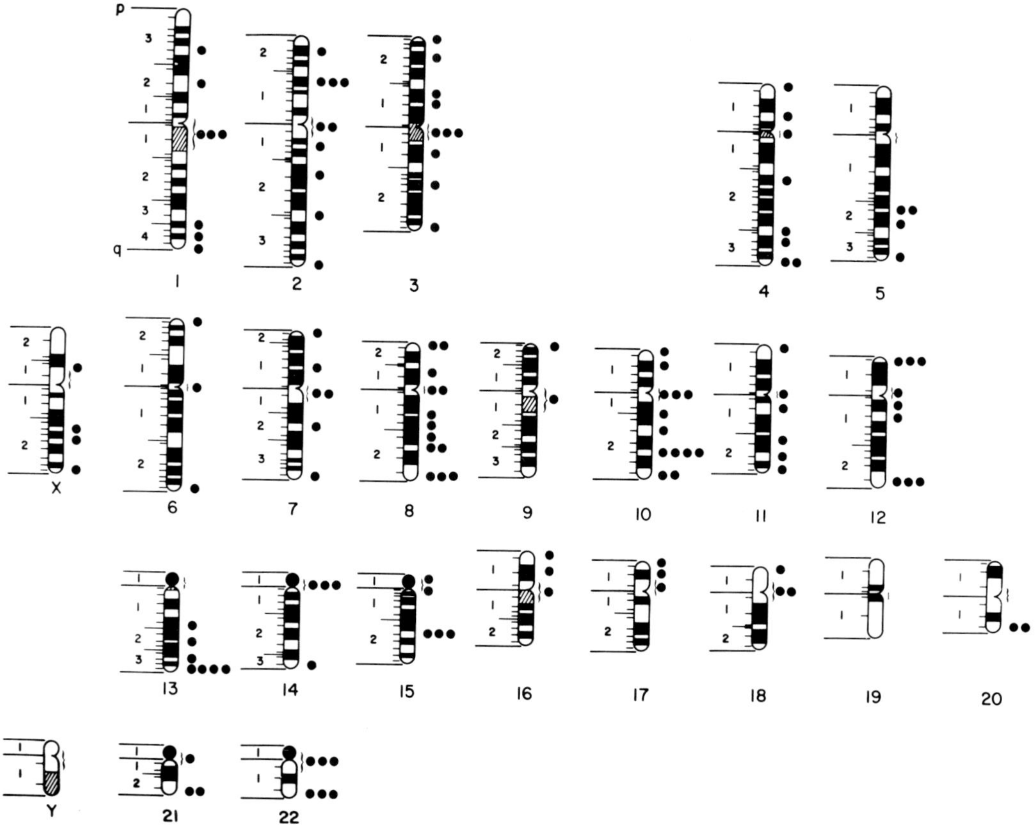

Fig. 7. a General distribution of all localizable breaks found in irradiated embryonic lung fibroblast cultures. Same experiment as the one illustrated in figure 6.

radiation on the doubling potential of human fibroblast populations and the frequency (%) of chromosome rearrangements observed in the respective nonirradiated cells through their in vitro life span. No chromosome rearrangements were observed in cells whose life span was shortened by radiation (skin fibroblasts from normal donor). By contrast, a high level of rearrangements was observed in the cell population whose doubling potential was increased by radiation (skin fibroblasts from a Fanconi's anemia patient). Finally, the cells whose life span was not affected by radiation (human embryonic lung fibroblasts) had an intermediate number of chromosome rearrangements. These results suggest

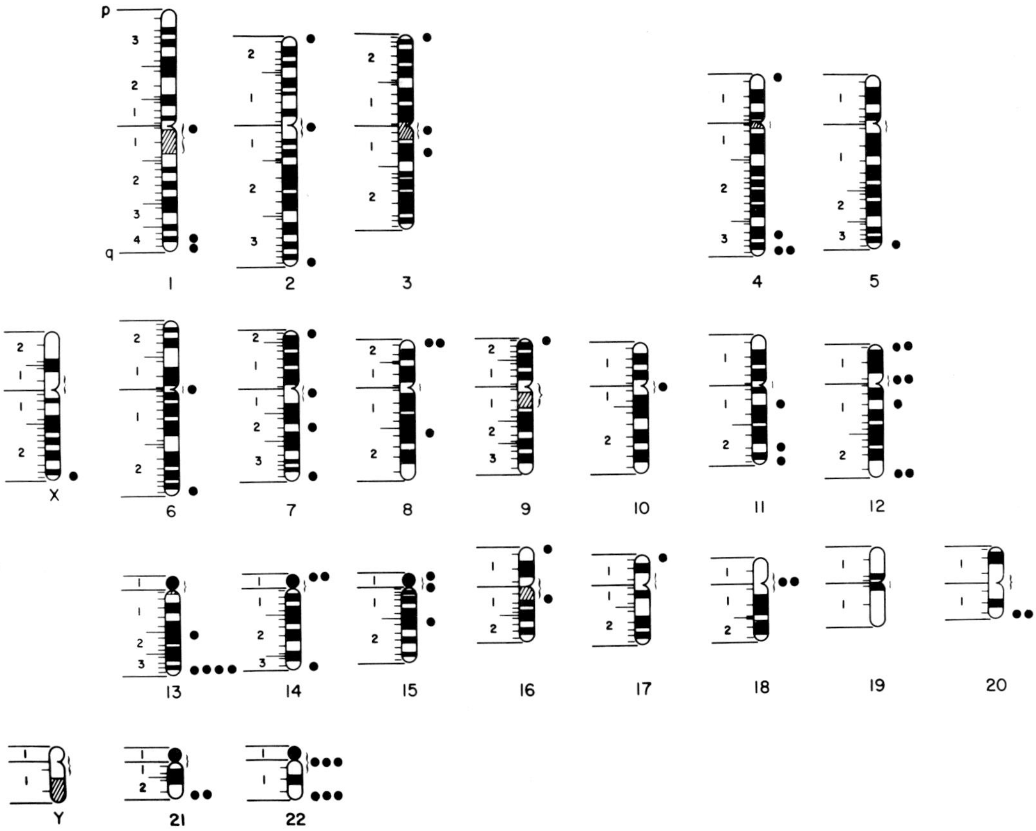

Fig. 7. b Distribution of breaks involved in exchanges [Bourgeois et al., 1981].

that the effect of radiation depends upon the potential of the cell populations for chromosome rearrangements and that ionizing radiation under the experimental conditions used (low dose rate) accentuates the intrinsic growth potential of fibroblasts maybe by accelerating changes in the genome which anyway take place during cell replication.

Another pertinent observation was the finding [Bourgeois et al., 1981] that in the irradiated cells most of the breaks involved in the exchanges (53 out of 62) concerned the centrometric and telomeric regions (fig. 7). Thus, the intrachromosomal break distribution where exchanges took place was preferentially located at regions rich in repeti-

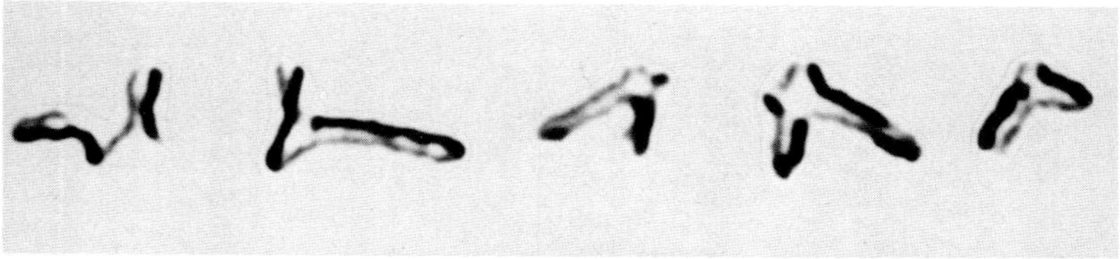

Fig. 8. Radial figures found in the chromosomes of mouse fibroblastic cultures before immortalization [Macieira-Coelho, 1980].

tive DNA, which has been implicated in recombinational events [Schmid and Jelinek, 1982].

Low dose rate ionizing radiation can also accelerate the acquisition of an infinite growth potential in mouse fibroblasts in vitro [Macieira-Coelho et al., 1976], a cell whose genome seems to be endowed with a high degree of 'plasticity'. Indeed, additional evidence in favor of an association between genome 'plasticity' and prolonged division potential comes from observations on some characteristics of the mouse cell genome. The chromosomes of mouse fibroblasts cultivated in vitro show a very unusual capacity of recombination; images of crossing over and bridges between chromosomes (fig. 8, 9), suggestive of interchromosomal exchange can be observed with a high frequency [Macieira-Coelho, 1980]. Another difference between the mouse and the human genome is the higher rate of sister chromatid exchanges in the former (table II) and the rapidity with which mouse cells can switch from the diploid to the tetraploid state (fig. 10). Furthermore, the chromatin lability which accumulates during cellular aging disappears in the case of mouse cells during these chromosomal rearrangements that occur during the transition from senescence to immortalization (fig. 11) [Beaupain et al., 1980]. We have also observed an interesting phenomenon of DNA elimination in mouse fibroblasts during the period preceding immortalization, which could be germane to the disappearance of the fragile chromatin sites. DNA measurements on cells in interphase after ethidium bromide staining showed that occasionally at early PDL the DNA content of G2 cells is less than that expected from the G1 content (fig. 12). This suggests that a process of DNA elimination goes on in these cells as has been reported for other organisms [Yao et al., 1984]. Although, as will be later described,

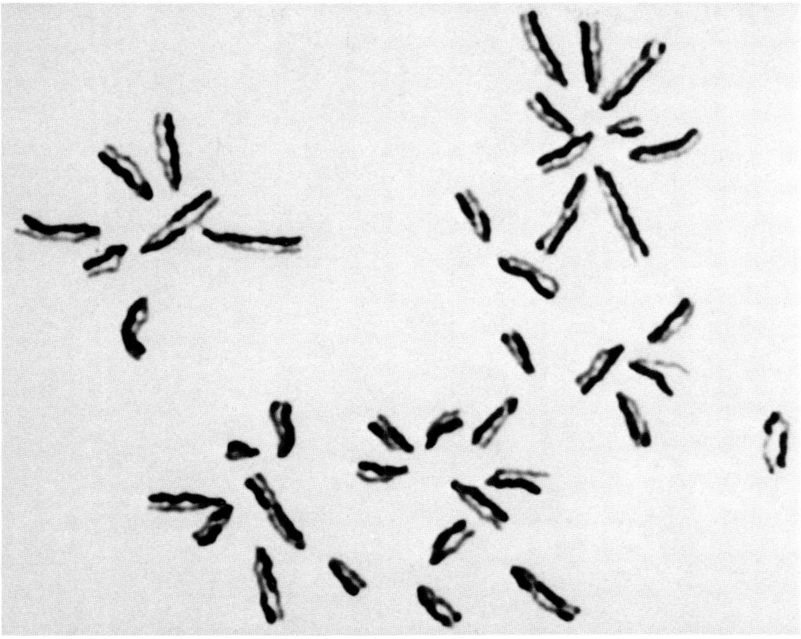

Fig. 9. Bridges between chromosomes found in mouse fibroblastic cultures before immortalization.

Table II. Number of SCE/cell[1] in human and mouse fibroblasts at different PDL after two division cycles in the presence of BrdU

Human		Mouse	
PDL	SCE	PDL	SCE
33	7	1	11
37	5	3	11
41	8	4	17
46	6	12	12

[1] Mean of 100 cells.

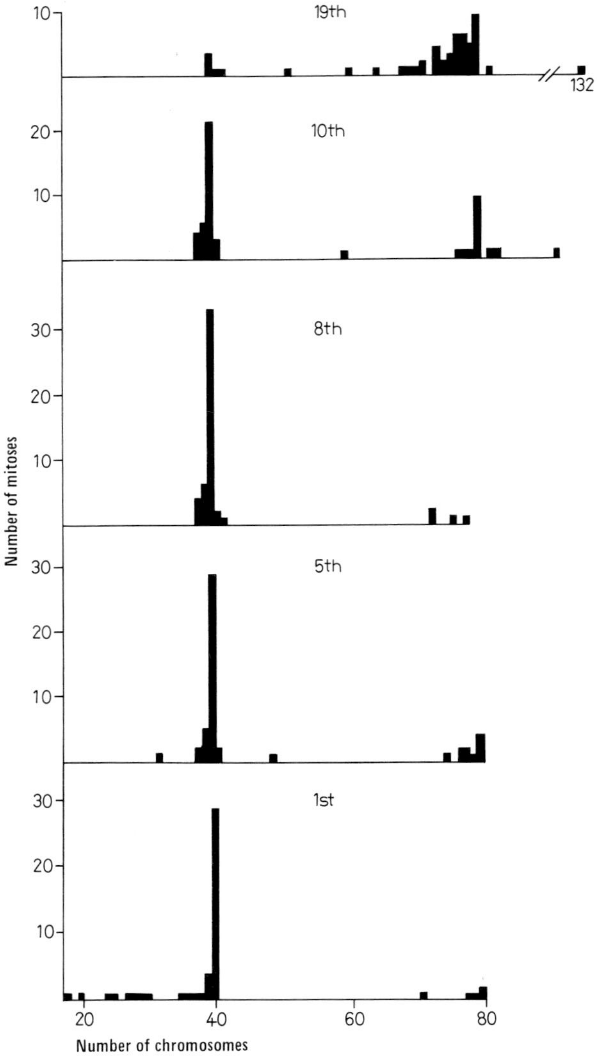

Fig. 10. Evolution of the ploidy at different population doubling levels of cultivated fibroblasts from newborn mouse lung. Between the 10th and 19th population doubling the number of diploid cells decreases and that of tetraploid cells increases significantly.

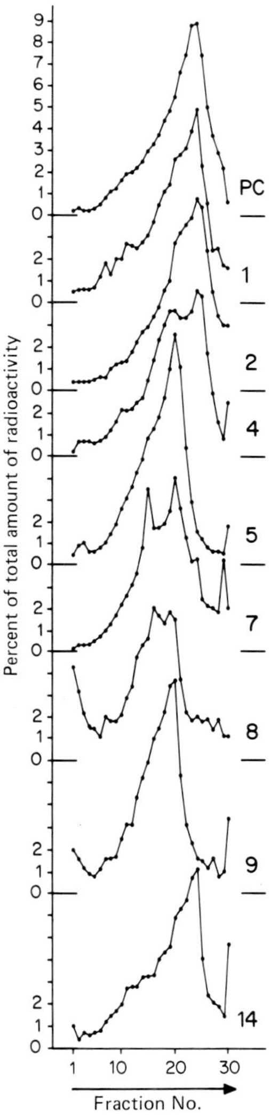

Fig. 11. Radioactivity found in the different fractions collected from alkaline sucrose gradients after centrifugation of chromosomal DNA from cultivated mouse fibroblasts at the indicated population doubling levels. PC = Primary culture. Smaller molecular weight DNA is present during the growth crisis [7–8 population doublings) and disappears during immortalization [Beaupain et al., 1980].

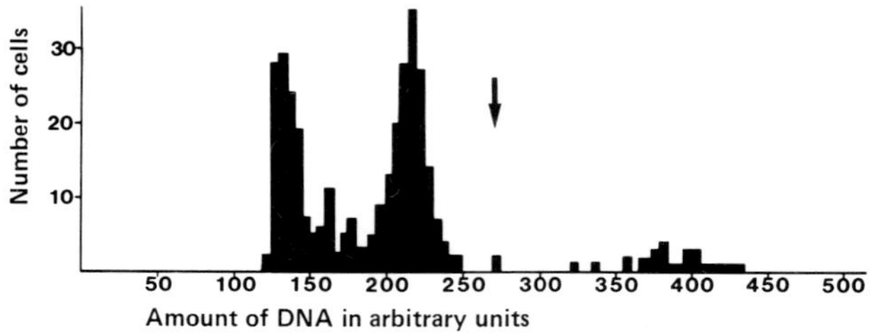

Fig. 12. Distribution of DNA contents in interphases of proliferating mouse fibroblast cultures before immortalization. The arrow indicates where one should find the peak of cells with G_2 DNA content as judged from the first peak corresponding to the G_1 cells.

DNA elimination also takes place during the division cycle of human cells, it is not of the same amplitude as in mouse cells.

This 'plasticity' of the mouse genome could well be at the basis of the propensity to escape senescence at the cellular level and to progress in the path of transformation. It is also probably due to this 'plasticity' that in the wild, different mouse subspecies are found with different chromosome sets [Capanna, 1983].

So, it is not the chromosome constitution that is important for cell longevity [Hoehn et al., 1980] but rather the capacity of the genetic material to recombine, as suggested from the data described above. Hence there are considerable data showing that the problem of the growth potential of cells in vitro is not an artefact of in vitro cultivation and has important correlations with evolutionary, developmental and physiopathological characteristics of the organism.

The observations on the relationship between the capacity of cells to divide and the potential for recombinational events of their genome, led to a new paradigm for the explanation of the long-term growth potential of cells in vitro [Macieira-Coelho, 1979, 1980, 1981b]. According to the paradigm, the trigger for a progressive shift in cell behavior leading to senescence is the genome reorganization which inevitably accompanies cell division and modifies DNA at different levels of its structure. This is due inter alia to chromosomal rearrangements, sister chromatid exchanges, DNA strand switching, DNA loss, displacement of transposable elements, gene amplification, deletions and other events that occur during the cell cycle. This reorganization could be the inevitable event which, although providing the cell with a way to evolve, adapt and survive, will also determine its permanent drift and lead to the loss of the division potential. In some cells, however, there is a higher probability to overcome this reorganization and to escape from senescence.

There is a considerable amount of evidence supporting the idea that a reorganization of the genome takes place during senescence of dividing cells. It remains to be found out what are the characteristics of certain genomes that allow them to overcome this reorganization and maintain its 'plasticity'.

Several phenomena show that indeed a genome reorganization takes place during the division cycle. One is the well-known sister chromatid exchanges [Taylor, 1958]; another is the asymmetry in sister chromatids after DNA replication that can be observed with chromosome banding [Goradia and Davis, 1978].

Direct evidence for recombination events taking place during serial divisions comes from cytogenetic analyses performed on human fibroblasts where chromosome markers were used to detect somatic segregation in cells obtained from subjects heterozygous for chromosomal variants. Somatic segregation (chromosome recombination) for three different autosomes could be observed [Martin and Sprague, 1969]. Recombinant diploid cells appeared within cultures of tetraploid clones with a cycle 2 n→4 n→2 n.

Additional evidence for genome reorganization during the cell cycle comes from cytogenetic experiments showing that several rearrangements take place during serial divisions of cells. Human fibroblasts have been well studied in this respect. Translocations, inversions, dicentrics, breaks and deletions have all been described and increase through the cell population life span [Saksela and Moorhead, 1963; Benn, 1976; Miller et al., 1977].

A pertinent observation is the one reported by Chen and Ruddle [1974], already cited, who found that during the proliferation of normal human embryonic fibroblasts, there are minor stem cell populations which compete, but none that dominate. The cells are not uniformly diploid; rather, several recurring chromosomal rearrangements can be detected at sufficient high frequencies to insure their generation by minor stem cell populations [Chen and Ruddle, 1974]. These authors could not detect systematic shifts in the banding patterns, suggesting that the chromosomes are structurally stable.

Miller et al. [1977] made a particularly thorough study of several cultures obtained from human kidney and lung. The kidney yielded epithelioid and the lung fibroblast-like lines. Significant increases in

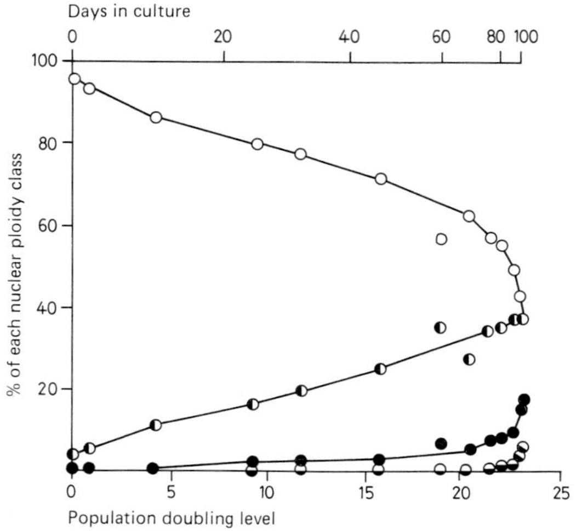

Fig. 13. Variations in the percentage of each nuclear ploidy class as a function of population doubling level (2 C ○—○; 4 C ◐—○; 8 C ●—●; 16 C ◐—◐) [Kaji and Matsuo, 1981].

polyploidy occurred during the earliest passages but then declined. There were low levels of cell aberrations among mid-passage cultures and during the last third of the life span there were increases in polyploidy and breakage type aberrations. This pattern of chromosomal changes through the life span of human embryonic fibroblasts is in agreement with the results of Matsuo et al. [1982]. The evolution of the ploidy was also followed in serially dividing chicken fibroblasts [Kaji and Matsuo, 1981]. The percent of diploid nuclei with a 2 C DNA content declined progressively, while the 4 C nuclei increased (fig. 13). Since close to 100% chicken cells synthesize DNA up to the end of the population life span [Lima and Macieira-Coelho, 1972] and the number of multinuclear cells increases steadily (fig. 14) [Kaji and Matsuo, 1978], in this species the disturbance in cytokinesis must precede the disturbance in DNA synthesis.

These rearrangements at the level of chromosomes must correspond only to the visible part of the iceberg and one may wonder what happens at other levels of DNA organization [Macieira-Coelho, 1984]. The changes could be of a quantitative and qualitative nature (fig. 15). The former could be due to loss or unequal distribution of DNA during cell division or to the integration of new DNA. Qualitative changes could be

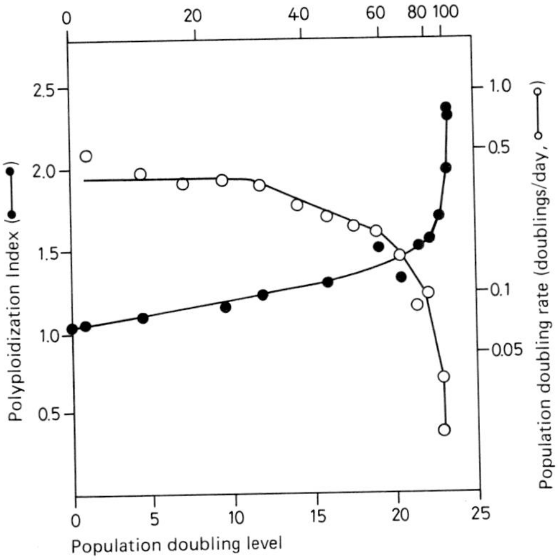

Fig. 14. Variations in polyploidization index and population doubling rate as a function of population doubling level [Kaji and Matsuo, 1981].

Genome reorganisation during cell division	Quantitative changes	Loss of DNA Unequal distribution of DNA between sister cells Integration of new DNA
	Qualitative changes	Methylation Chromatin conformation Mutual relationship between chromatin domains coming together or being separated Transposable elements Unequal recombination Breaks Loss of DNA Integration of new DNA

Fig. 15. Schematic representation of the possible events implicated in the reorganization of the genome during cell division.

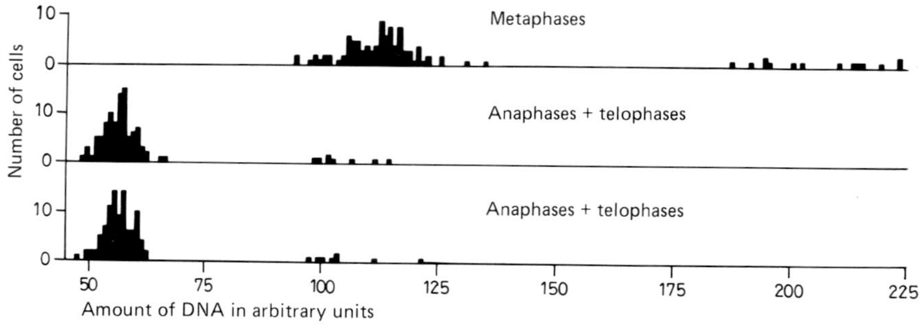

Fig. 16. Distribution of DNA content found on metaphases and on anaphases and telophases of human embryonic lung fibroblasts. Ethidium bromide staining [Macieira-Coelho et al., 1982].

caused inter alia by changes in chromatin structure and conformation, by the mutual relationship between chromatin domains coming together, by the transposition of controlling elements, by breaks, by the loss of DNA or integration of new DNA, by DNA methylation, by gene amplification and gene conversion.

The problem of quantitative changes in DNA through cell division, was first approached with cytofluorometry and cytophotometry [Macieira-Coelho et al., 1982]. DNA measurements were made on cells in interphase, prophase, metaphase and anaphase. It was found that the scatter of DNA contents in different cells is the same regardless of the stage of the cell cycle.

To see if the scatter of DNA contents between cells varied with cell divisions, the distribution of the DNA contents of each half of anaphases and telophases, and that of metaphases (fig. 16) from cultures stained with ethidium bromide at different population doubling levels was plotted on probit paper in such a way that the scale of the abscissa for metaphases was twice that of anaphases plus telophases (fig. 17). The values corresponding to each half of anaphases and telophases and to metaphases overlapped and their slopes were never significantly different. This reinforced the idea that the variation of DNA content between cells is due to quantitative differences and not to within-sample variation caused inter alia by changes in the binding of the dye to DNA at different stages of the division cycle or to a different geometry of the cell. Staining with Feulgen also showed that the same DNA contents are obtained regardless of the stage of the cell cycle (fig. 18).

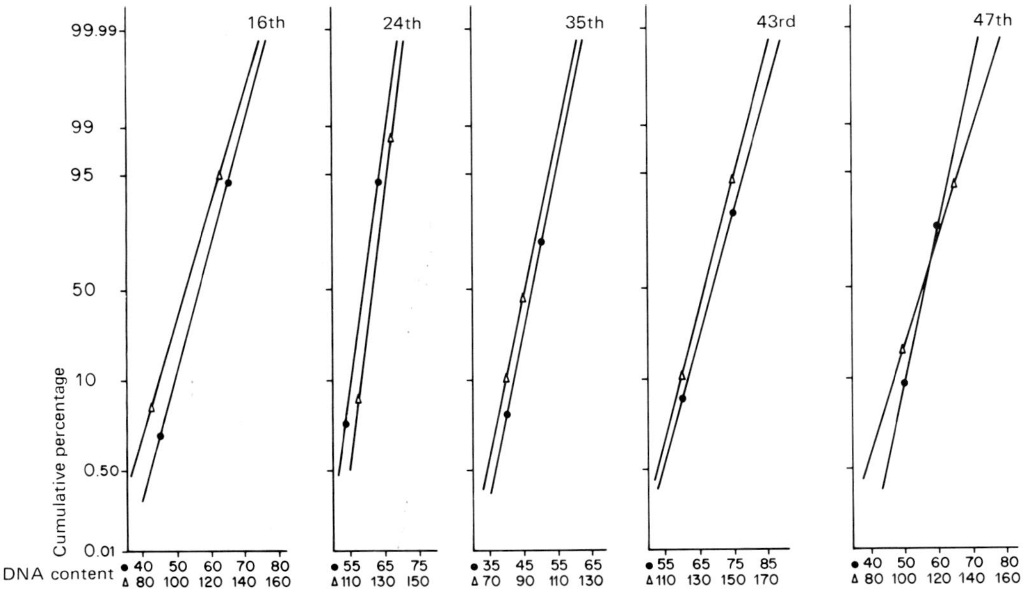

Fig. 17. Regression lines, computed by the least-squares method, of the distributions of DNA content (ethidium bromide staining) found on metaphases (△) and on anaphases and telophases (●) of human embryonic lung fibroblasts at different population doubling levels. The scale for metaphases is double that for anaphases and telophases [Macieira-Coelho et al., 1982].

The DNA content of each half of anaphases and telophases was also compared to see if the differences in DNA contents originate at the time of separation of the daughter cells. The differences were expressed as the percentage of the mean value of each pair; the data showed that in at least 20% of the cells, differences between sister cells are significant and not due to methodological errors (fig. 19). The fraction of cells at each doubling, with uneven distribution of DNA between daughter cells, was constant throughout the life span of the fibroblast population and increased only during the last 2–3 doublings (fig. 20).

The spread of the distribution of DNA contents did not vary through the cell population life span (fig. 21) although in terminal cells there seemed to be two different modes. To determine whether the differences in the DNA content between cells were caused by unequal distribution of the DNA synthesized during the preceding S period, the number of grains on each half of anaphases and telophases was measured after label-

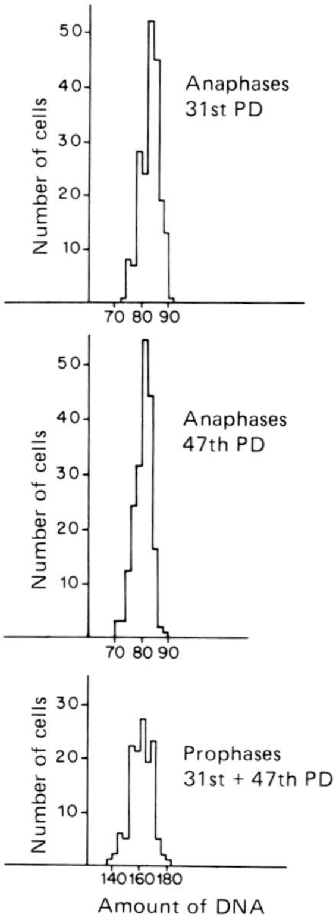

Fig. 18. Distribution of DNA content (Feulgen-pararosaniline staining) found on anaphases and prophases of human embryonic lung fibroblasts at the 31st and 47th population doubling.

ing the DNA with ^{3}H-thymidine during a whole S period. To chose the amount of the precursor that does not interfere with cell division, cells were labeled with different concentrations of ^{3}H-thymidine and the distribution of the differences in DNA contents between anaphases and telophases found in the different groups of cultures was plotted (fig. 22). It was found that beyond 0.2 μCi/ml there was a decrease in the number of pairs with less than 4% differences and an increase in the fraction of cells with larger differences. Thus, radioactivity increased the fraction of cells with significant differences in DNA contents. This was considered as proof that beyond 4% of the mean of the pair, the differences in DNA

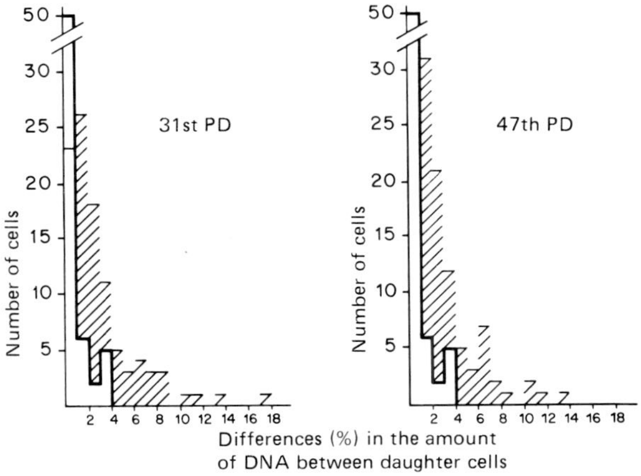

Fig. 19. Distribution of the differences in the amounts of DNA (Feulgen-pararosaniline) expressed as percentage of the mean of the pair, found between daughter cells, in human embryonic lung fibroblasts at the 31st and 47th population doubling. The area limited by the continuous line corresponds to the differences with three standard deviations, found between repeated measurements of the same samples [Macieira-Coelho et al., 1982].

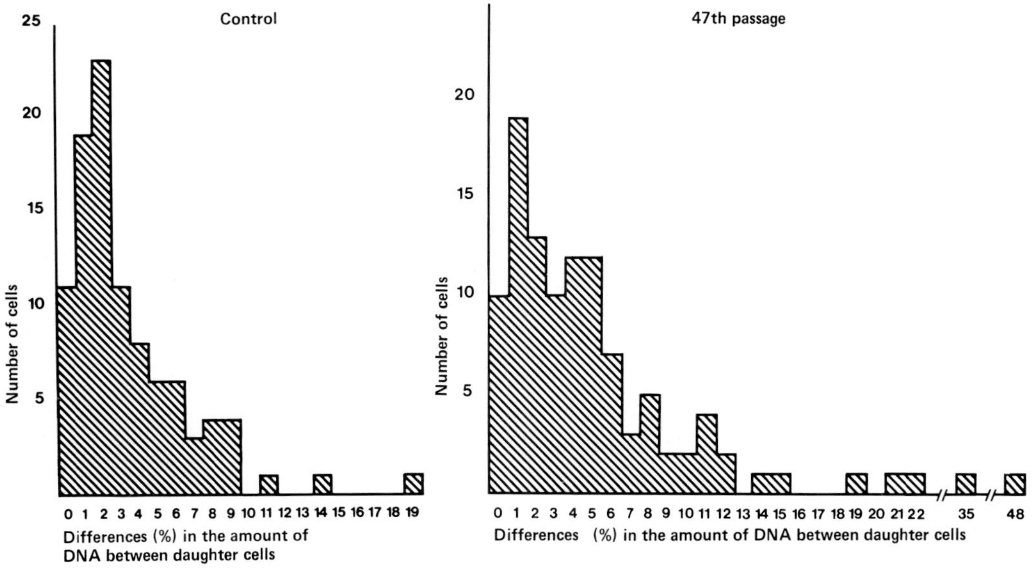

Fig. 20. Distribution of the differences in the amount of DNA (ethidium bromide staining) expressed as percentage of the mean of the pair, found between daughter cells in phase II (control) and phase IV (47th passage) human embryonic lung fibroblasts [Macieira-Coelho et al., 1982].

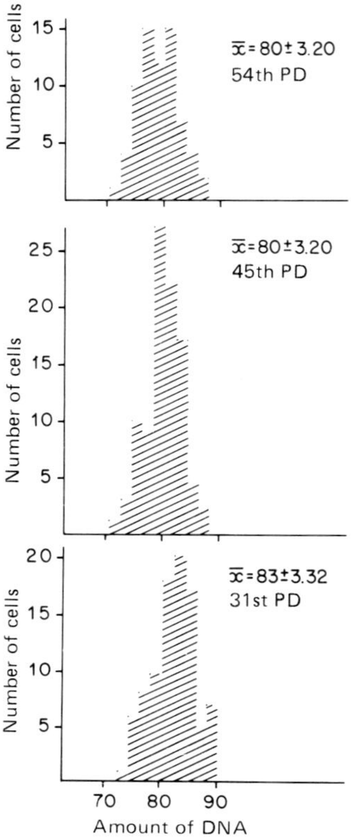

Fig. 21. Distribution of DNA content in arbitrary units found on anaphases of human embryonic fibroblasts at different doubling levels, stained with Feulgen-pararosaniline. The mean DNA content with the 95% confidence limits, at the respective population doubling, is indicated [Pouvion-Dutilleul et al., 1984].

contents between sister cells are significant [Macieira-Coelho et al., 1982].

Nontoxic concentrations of ^{3}H-thymidine were then used to measure the distribution of grains on each half of anaphases and telophases, on autoradiographs of cells labeled during a whole S period. The plot obtained with this experiment was compared with a computer simulation, which showed that in a significant fraction of cells that did not vary throughout the population life span, newly synthesized DNA was not distributed evenly between daughter cells [Macieira-Coelho et al., 1982]. The fraction of cells with significant differences was smaller than that found

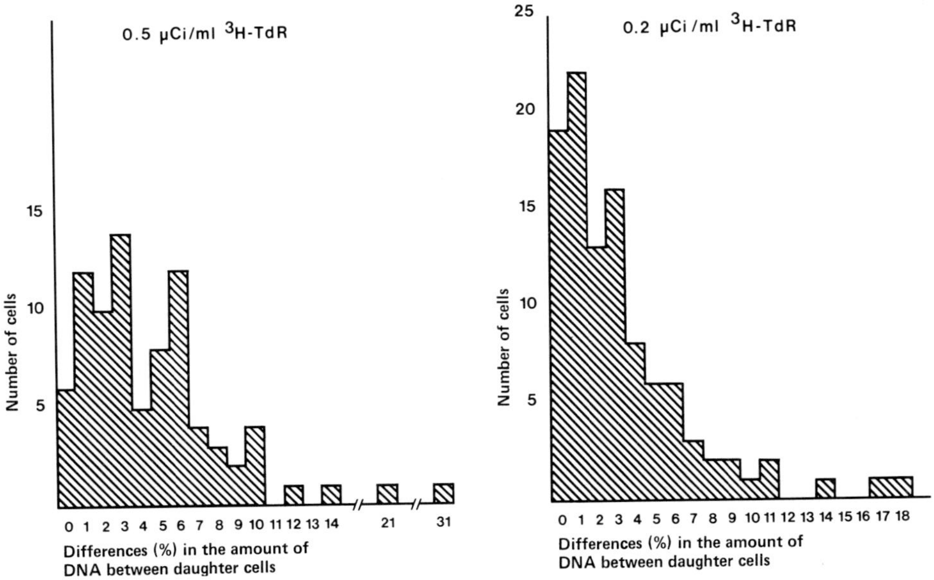

Fig. 22. Distribution of the differences in the amount of DNA (ethidium bromide staining) expressed as percentage of the mean of the pair, found between daughter cells in cultures labeled with 0.2 and 0.5 μCi/ml ^{3}H-TdR [Macieira-Coelho, 1982].

measuring DNA contents, which was to be expected since the precursor is incorporated in a limited number of sites, while cytophotometry on the contrary evaluates the whole DNA content.

The quantitative changes in DNA could originate during semiconservative DNA synthesis, chromosome assembly, or chromosome segregation. It was concluded that the three mechanisms are implicated [Macieira-Coelho et al., 1982]. Indeed, when screening for mitoses, disturbances in chromosome segregation can be detected (fig. 23).

These results showed that this process of uneven distribution of DNA, goes on at a constant rate in serially dividing cells and becomes amplified only at the very end. It must be a source of genetic drift and of progressive cell heterogeneity.

Results obtained with lymphocytes from human donors of different ages, showing an increased variance with age in the DNA content between cells [Staiano-Coico et al., 1982] could be germane to the results obtained in vitro by Macieira-Coelho et al. [1982].

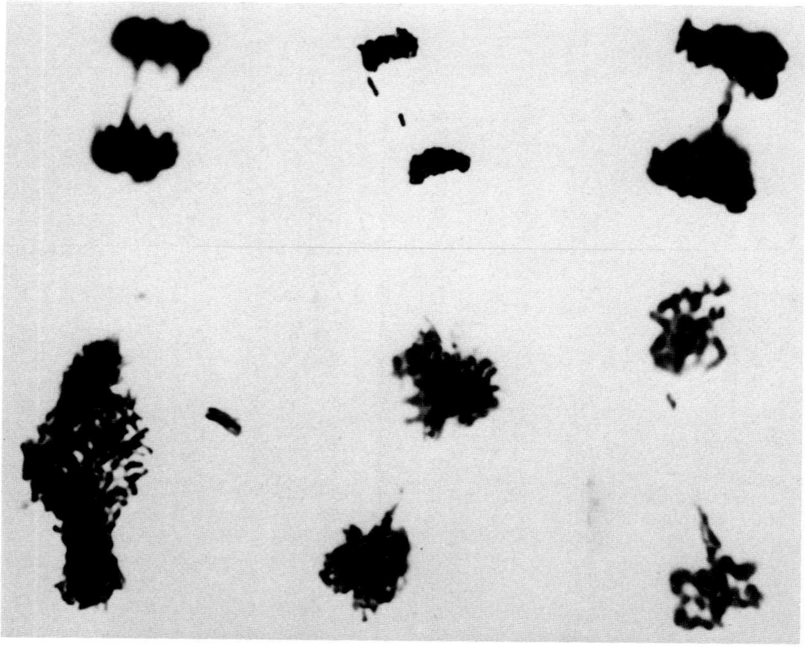

Fig. 23. Loss of chromosomal material and defective chromosomal segregation observed during anaphase in human embryonic lung fibroblast cultures.

In regard to the dogma of the semiconservative partition of DNA between sister cells, one can now say that the events are much more complex when analysed at the level of the individual cell. In fact, the genetic information received by the daughter cell is different from that of the other daughter cell and from that of the mother cell.

The problem of searching for qualitative differences in the high order structure of DNA was first approached studying the organization of the chromatin at the level of the 30-nm solenoid at different population doublings of human embryonic fibroblasts [Puvion-Dutilleul and Macieira-Coelho, 1982]. The density and the spacing of the fibers at the periphery of the nuclei were measured by image analysis (fig. 24) [Macieira-Coelho et al., 1987]. The density of the fibers, mainly at the level of their anchorage to the lamina densa, was found to decrease progressively (fig. 25). On the other hand, the spacing of the fibers was found to follow a two-step pattern. The fall in the density correlated with the decline in the

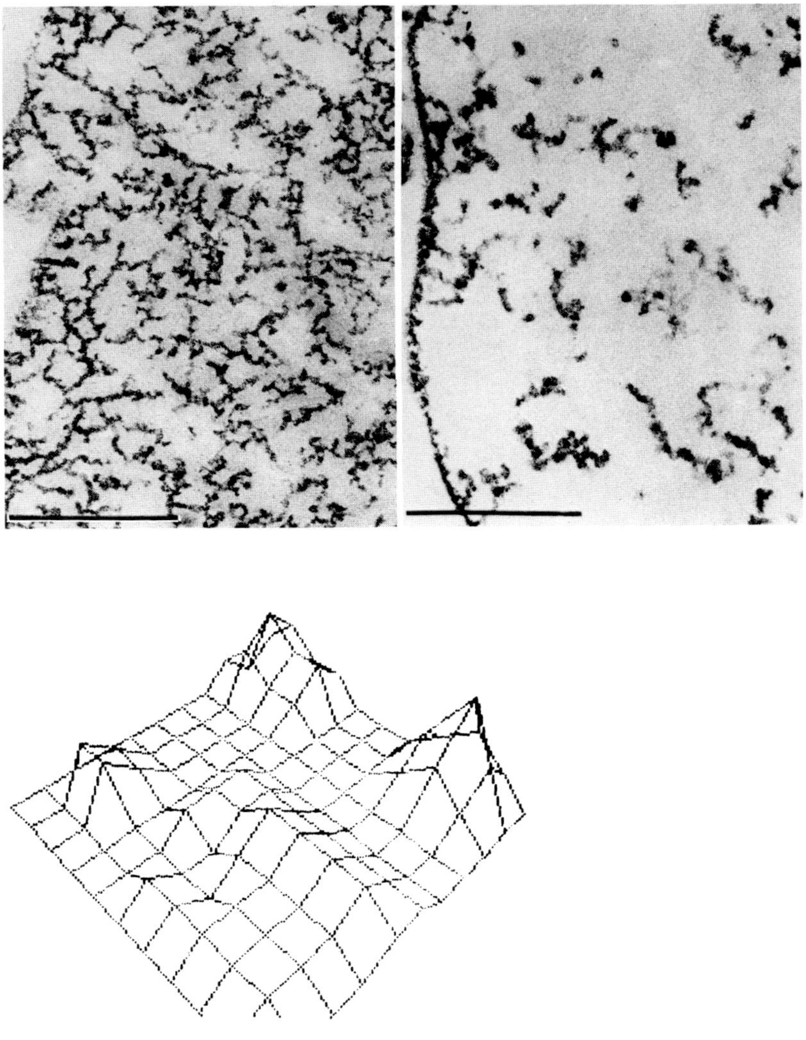

Fig. 24. Periphery of young (top left) and old (top right) nuclei of human embryonic fibroblasts prepared for the visualisation of the 30-nm chromatin fibers [Puvion-Dutilleul and Macieira-Coelho, 1982]. Bottom: Computer-assisted morphological transformation obtained by screening pictures identical to those shown above.

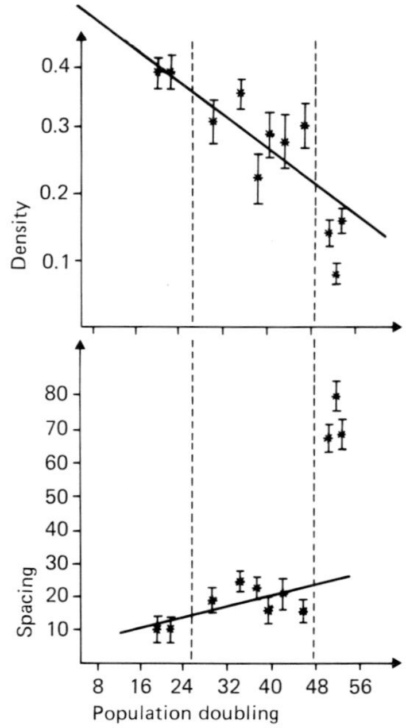

Fig. 25. Mean density and mean spacing at different population doubling levels, of the 30-nm chromatin fibers from serially cultivated human embryonic lung fibroblasts, obtained by image analysis. The 95% confidence limits are indicated by the vertical bars. The vertical dashed lines indicate the approximate limits of phases II, III and IV, determined by the analysis of the kinetics of proliferation.

rate of DNA synthesis initiation after cell attachment and the increase in spacing coincided with the fall in the maximal % of cells capable of synthesizing DNA during a 24-hour period before resting phase is reached (fig. 26). These results support the view that conformational changes take place in chromatin; they are probably caused by structural modifications both of chromatin and of the scaffold where the latter is anchored.

Additional evidence for genome reorganization at another level of DNA structure came from the finding of an increased heterogeneity in DNA repeat lengths during serial divisions of human postnatal skin fibroblasts [Dell'Orco et al., 1986] (fig. 27). This change in the rate of DNA digestion by micrococal nuclease could also be due to conformational changes occurring during cell senescence.

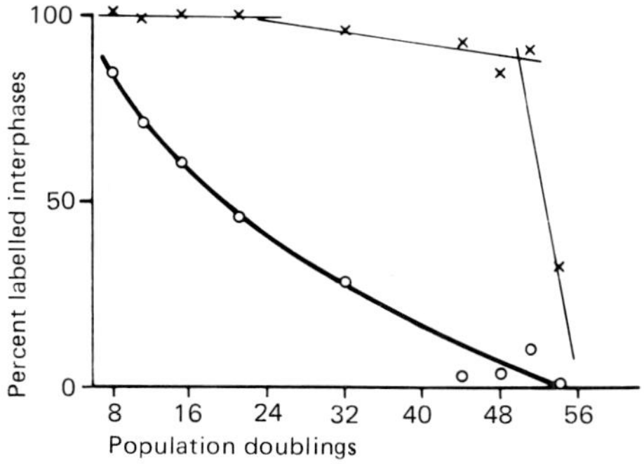

Fig. 26. Percent labeled interphases found during the first 24 h after subcultivation (○) and maximal percentage reached during a 24-hour period between subcultivation and resting phase (x) in cultures subcultivated with a 1:16 split at different population doubling levels. The lines are roughly drawn [Macieira-Coelho and Azzarone, 1982].

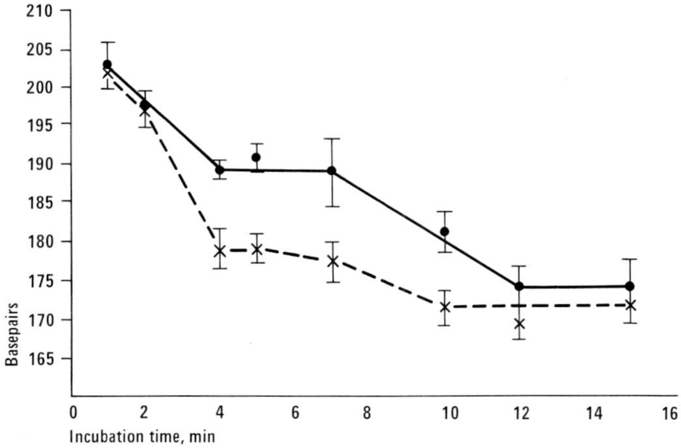

Fig. 27. DNA repeat lengths in base pairs were determined from low population doubling level, 6 experiments PDL 13–23 (●—●, mean ± SEM) and from high population doubling level cells, 7 experiments 51–79 (x----x; mean ± SEM), after digestion with micrococal nuclease from 1 to 15 min [Dell'Orco et al., 1986].

Reorganizations at the gene level are suggested from the decreased hybridizability of the probes for a specific family of repeated sequences [Shmookler Reis and Goldstein, 1980] and of the probes for some genes, though not for all. The latter, which concerned the alpha-globin and beta-actin genes, was observed during serial population doublings of embryonic fibroblasts [Icard-Liepkalns et al., 1986] and of postnatal skin fibroblasts from normal human donors [Icard-Liepkalns et al., unpubl.]. No changes could be observed with the beta-interferon gene [Icard-Liepkalns et al., 1986] and with the H3 and H4 histone genes [Green et al., 1986].

As will be described in the events occurring during phase IV, the decreased hybridization to the alpha-globin and beta-actin genes must be due in part to loss of those genes into extrachromosomal circular DNA. This, however, cannot be the sole explanation since the decline in the hybridization signals occurs well before the increase of DNA in an extrachromosomal form, an event that seems to occur only during the transition to the postmitotic state [Icard-Lipkalns et al., 1986].

A decreased hybridizability of the DNA coding for ribosomal RNA has been previously reported in vivo during aging of mammals [Johnson and Strehler, 1972; Strehler et al., 1977; Strehler and Chang, 1979] and in *Tetrahymena thermophila* [Allen et al., 1984]. In the latter, the losses of genes could be attributed to deletions of particular chromosomes. Several other mechanisms, besides gene loss, could, however, be implicated in the decreased hybridizability, such as mispairing deletions [Strehler, 1978] and an age-related increase in tightly bound proteins to DNA [Gaubatz and Cutler, 1978]. Indeed, Gaubatz and Cutler could demonstrate a recovery of the hybridization of the probe with genomic DNA, after intense deproteinization of the DNA. Other possible reasons for a decreased hybridizability are protein-DNA cross-links and DNA-DNA cross-links. Evidence in favor of increased cross-links due to oxidized SH groups, during in vivo aging, has been obtained [Tas et al., 1980].

The amplification of the C-Has-ras gene in human fibroblasts aged in vitro [Srivastava et al., 1985] also favors the idea of a genome reorganization during aging of dividing cells.

Icard et al. [1979] have reported a progressive increase in alkaline labile sites during serial proliferation of human embryonic fibroblasts in vitro. Subsequently, it was found that these lesions were due to an increased thermolability and alkaline lability of chromatin [Puvion-Dutilleul et al., 1984]. In the former work, circular pieces were cut with a drill from the bottom of the Petri dish where the cells were grown, and then layed in a lysing solution on top of a sucrose gradient. This method caused heating of the cells. Temperature alone was not enough to create double strand breaks, since when the same procedure was used and the chromosomal DNA was centrifuged in a neutral gradient, no breaks could be detected [Icard et al., 1979]. So, during phases I, II and III no breaks seem to be present in the living cell, which agrees with the experiments of Mayer et al. [1986]. Hornsby and Harris [1987] could not observe either, in adrenocortical cells, an accumulation of DNA damage. In human fibroblasts, what seems to take place is an increased fragility of chromatin due probably to conformational changes and to modifications of its anchorage on the nuclear cage. The progressive increase in cell volume and the loss of cell motility must be due to a loss of the plasticity of the cellular scaffold with repercussions on the molecular structures that the latter supports. Indeed, the visualisation of chromatin at the ultrastructural level also suggests that the whole nucleus becomes more sensitive to shear forces [Puvion-Dutilleul et al., 1982, 1984]. Only at the very end (phase IV) is there a sudden increase in small molecular weight DNA in parallel with other events that precipitate during the last doublings. These will be treated in more detail in another section.

Other changes in DNA concern methylation, a process known to be necessary for variations in the handedness of DNA. It was found that methylation drifts in a random fashion, in particular for repressed genes, which could lead to random derepression [Shmookler Reis and Goldstein, 1982a, b]. On the other hand, the methylation of three highly repeated DNA families remained relatively constant. The shortening of the cell division potential by drugs known to inhibit the methylation of cytosine was taken as evidence supporting the view that random loss of methyl groups in DNA may also be responsible for the loss of the division potential [Holliday, 1986; Fairweather et al., 1987].

With such an obvious remodeling going on in the genome through serial divisions, it is not surprising that progressive modifications can be detected at different cellular levels. Some are directly or indirectly related with the commitment to divide, others may have other implications for cell physiology.

2. Search for Nuclear and Cytoplasmic Determinants of the Growth Potential

a. Fusion Products between Cells with Different Growth Potentials

A particular fruitful series of experiments were those attempting to reinitiate DNA synthesis in terminal phase IV cells through fusions with actively replicating cells. The first experiment of this type achieved the rescue of the old human cells through hybridization with an immortal hamster cell line, but the methodology used did not exclude the mating with residual replicating cells [Goldstein and Lin, 1971].

More conclusive experiments were performed by Norwood et al. [1974] measuring the DNA synthetic activity in heterokaryons of young and old nuclei. They showed that the senescent nucleus inhibited DNA synthesis in the young nucleus. The same group subsequently found [Norwood et al., 1975], with the same approach, that in heterokaryons with nuclei of old human fibroblasts and of HeLa cells or SV_{40}-transformed human fibroblasts, DNA synthesis was reinitiated in the old nucleus. HeLa cell cytoplasm could reinitiate DNA synthesis in only 20% of terminal normal fibroblast nuclei, as compared to the induction when the HeLa nucleus was present. Whole cells, karyoplasts and cytoplasts from the established L mouse cell line, could also stimulate DNA synthesis in nuclei from fibroblasts from an 82-year-old donor [Nette et al., 1982].

Nuclei from growth-arrested young cells were inhibitory for nuclei from replicative young cells [Rabinovitch and Norwood, 1980; Stein and Yanishevsky, 1981] if fusion occurred at least 3 h before the G1/S boundary. It was suggested that the mechanism of arrest due to aging and that due to contact inhibition of growth in young cells share a final common pathway. It is reasonable to assume that the same inhibitor caused the growth arrest during cell crowding and in the terminal stage; however, other metabolic events, as we later describe herein, suggest that the terminal cell corresponds to a distinct event. It was further shown that cells in S phase at the time of fusion, complete DNA synthesis [Stein and Yanishevsky, 1979]. On the other hand, young replicative cells induce DNA synthesis in young cells showing that the initiation of the synthesis in the latter is positively controlled [Stein, 1983].

Nuclei from young fibroblasts exposed to sublethal doses of amino acid analogs or to mitomycin C behaved like old nuclei in the inducement and repression of DNA synthesis suggesting that cells which leave the proliferative pool as a result of a variety of mechanisms may behave similarly in heterokaryons [Norwood et al., 1979].

Brief postfusion treatments with cycloheximide or puramycin delayed the inhibition of DNA synthesis in heterokaryons with young and old nuclei [Burmer et al., 1982]. It suggested that a protein synthesis dependent growth inhibition is responsible for the supressive effect on DNA synthesis.

Burmer et al. [1983] and Drescher-Lincoln and Smith [1983] also showed that cytoplasts from senescent cells are inhibitory when fused with whole cycling fibroblasts. Perfusion treatment of the old cells with protein synthesis inhibitors eliminated the ability of the senescent cytoplasts to block DNA synthesis [Drescher-Lincoln and Smith, 1984]. The DNA synthesis inhibitory activity is associated with the cytoplasmic membrane, and is heat, trypsin and periodate sensitive suggesting that it is a glycoprotein [Stein and Atkins, 1986; Stein et al., 1986].

Furthermore, polyadenylated RNA isolated from old human fibroblasts inhibited DNA synthesis in young cells after microinjection, whereas the RNA from proliferation competent cells had no such effect [Lumpkin et al., 1986a]. Polyadenylated RNA from young, growth-inhibited cells had a slight inhibitory effect. These data strongly suggest that a protein inhibitor of DNA synthesis is present in quiescent cells and in increased amounts in terminal cells.

Stein and Yanishevsky [1979, 1981] and Stein et al. [1982] obtained heterokaryons with nuclei from terminal human fibroblasts and from different immortalized cell lines. They found that DNA synthesis was induced in the old nucleus when the other nucleus originated from SV_{40}-transformed or adenovirus-5-transformed human cells and from HeLa cells. However, when the other nuclei originated from a human tumor cell line, a chemically transformed human cell line, gamma-radiation and UV transformed human fibroblasts or from RSV-transformed cells, DNA synthesis was inhibited by the old nucleus. This implicates different mechanisms of transformation in these immortalised cell lines. It was suggested that the latter cell lines whose nuclei were inhibited have lost a regulatory factor but are still sensitive to the inhibitor, whereas the former have gained a dominant transformed factor capable of overriding the inhibitor of DNA synthesis and have lost the sensitivity to the inhibitor. Other factors are also implicated in the rescuing of phase IV cells, since the capacity of the latter to be rescued is correlated with the levels of DNA polymerase-alpha in the transformed cells [Pendergrasse et al., 1982]. Only cell lines with higher levels of the enzyme could induce DNA synthesis in old nuclei.

Infection with DNA viruses such as human cytomegalovirus and SV_{40} can also induce DNA synthesis in the terminal cell [Ide et al., 1983, 1984a]. Infection with SV_{40} virus, however, cannot induce cell proliferation [Matsumura et al., 1980]. The more rapid induction of proliferation and immortalisation of terminal cells by SV_{40} alone, previously reported

[Girardi et al., 1965], must be a rare event. An efficient immortalization though, can be achieved, if the cells are transfected both with the SV_{40} early region encoding the T antigen and with the xantine-guanine phosphoribosyl transferase gene [Mayne et al., 1986]. The need for these two markers shows that several events must take place in the genome for the cells to immortalize. Indeed, both the transfection and the acquisition of resistance to mycophenolic acid and aminopterin must be accompanied by several changes in the genome.

These series of experiments contributed to the understanding of the mechanisms leading to the inhibition of DNA synthesis in the terminal cell since they identified one of the causes of growth arrest, i.e. the production of an inhibitor. The long-term potential of cells, however, although obviously depending also on the ability to replicate DNA, is a much more complex phenomenon. This problem has been extensively studied by different groups.

An attempt was made to find out whether the control of the potential to divide is of nuclear or cytoplasmic origin, fusing anucleated cytoplasms from young and old cells to whole fibroblasts [Wright and Hayflick, 1975]. These experiments were made with human embryonic lung cells (WI-38) and compared hybrids between yound and old cells, with those made between young/young or old/old controls. Although the results favored nuclear rather than cytoplasmic factors, it was difficult to ascertain one or the other because of the harsh treatment to which the cells were submitted which also hampered the growth of the controls. Later, Muggleton-Harris and Hayflick [1976] with another methodology, separated cytoplasts and karyoplasts from young and old fibroblasts (WI-38) and reconstructed the whole cell. They could conclude that both the old nucleus could limit the number of doublings of a young cytoplasm and the old cytoplasm could shorten the life span of a young nucleus.

Hoehn et al. [1978] compared the replicative life spans of hybrids derived from short-lived and long-lived human skin fibroblast cultures and found that they had growth potentials intermediate to that of the parents. The same conclusion was reached by Muggleton-Harris and Aroian [1982] with crosses between replicating skin fibroblasts from human donors of different ages. Pereira-Smith and Smith [1982] performed the same type of experiment but with the older parent line close to the terminal stage. They found that the proliferative potential of the hybrids resembled that of the older parent, i.e. the phenotype of limited division was dominant in the hybrids.

Lipman and Muggleton-Harris [1982], also with the same rationale, fused normal with cataractus mouse lens epithelial cells, the latter having a reduced growth potential. In these experiments the hybrid cells had a mode of replication similar to that of the normal parental cells. It shows that the shorter division potential of the cataract cells can be corrected and, thus, is of a different nature from that caused by cellular senescence.

On the contrary, hybrids obtained from fusions of normal and Werner syndrome cell populations displayed the reduced growth potential of Werner cells [Salk et al., 1981b], suggesting that the shorter division potential of the fibroblasts from these patients is of the senescent type.

Muggleton-Harris and DeSimone [1980] studied the replicative potentials of fusion products between WI-38 cells and identical cultures transformed by SV_{40}. The replicative potential of the hybrids was similar to the control WI-38 cells in the majority of cases. Only less than 2% of the hybrids had a prolonged life span and possibly immortalized. Fusions between normal WI-38 cytoplasts and transformed karyoplasts did not show sustained replication. Mass culture hybridizations between Lesch Nyhan cells with finite life span (deficient in hypoxanthine-phosphoribosyl transferase) and HeLa cells showed that most hybrids had a limited life span slightly longer than Lesch Nyhan fibroblasts; only 1 in 10^5 cells were capable of indefinite replication. Fusions of Lesch Nyhan cells with HeLa cytoplasts did not show transformation [Bunn and Tarrant, 1980]. Pereira-Smith and Smith [1981] found that the majority of hybrids (70%) obtained between normal and SV_{40}-transformed human fibroblasts did not make more than seven population doublings; the others had a more extensive proliferation but a finite life span. Thus, in these hybrids, senescence could only be partially reversed. Only 1 or 2 foci/10^5 cells had indefinite growth. All hybrid clones including those that exhibited a finite life span, expressed the viral T antigen. In the hybrids, the SV_{40} genome was integrated in the same region as in the parental SV_{40}-transformed cells [Pereira-Smith and Smith, 1981].

Pereira-Smith and Smith [1983] obtained hybrids between normal human fibroblasts and immortalized cell lines that either can induce DNA synthesis in heterokaryons with old nuclei or whose nuclei are inhibited by the nucleus of the postmitotic cell. A higher percentage of the hybrids with the former cell lines was capable of dividing more than eight times. Fusion with the other cells, those that are inhibited by the old nucleus, resulted in fewer than 10% of the hybrids capable of more than 8 doublings. It was concluded that cellular immortality has a recessive nature. Surprisingly, in a number of fusions between immortal cell lines, the hybrids had a finite proliferative potential [Pereira-Smith and Smith, 1983].

These results have to be interpreted having in mind the data obtained by Martinez et al. [1978] showing that HeLa cell clones display a nonproliferative and a proliferative pool. The fastest growing clones and subclones segregate daughter cells with different growth potentials, showing a shift in the distributions toward lower growth rates and increase in the proportion of cells belonging to the nonproliferative pool. This phenomenon was called clonal attenuation. The same phenomenon exists in normal human fibroblast populations, but whereas the proliferating fraction disappears progressively in these cultures, in HeLa cells the two distinct pools are continually produced.

Hence in the fusions described above between immortalized and normal cells, some of the hybrids may have been obtained from cells committed to clonal attenuation. Furthermore, if HeLa cells display clonal attenuation, all other immortalized cell populations must have it to different degrees. It is possible that the different life spans obtained with the hybrids depending on the type of immortalized parental cell line, depends on the degree of clonal attenuation displayed by the latter. On the other hand, the methods used for fusion may cause genetic changes which accentuate the phenomenon of clonal attenuation in the hybrids.

It should also be kept in mind that the product of fusion between two nuclei is unpredictable and that during serial divisions the remodeling going on in these genomes is probably random. Furthermore, immortalization is a complex genetic phenomenon which can not be due to single genes and cannot be interpreted in terms of Mendelian genetics.

Le Borgne et al. [1978] obtained results which are representative of the complexity of the genetic effects triggered by hybridizations between eukaryotic cells. Clones were obtained between normal fibroblasts and immortal cell populations and between immortal cell populations from human and mouse origin. Contact inhibition of growth was measured in the parental lines and in the hybrids. Hybrids between normal and transformed human and between normal and transformed mouse cells were not contact inhibited. Fusion products between normal human and transformed mouse cells and between transformed human and mouse cells were contact inhibited. Hence, the results suggested that the inhibition of cell division due to crowding in the hybrids was more dependent upon the species of the parental cell lines rather than on their normal or transformed character. In other words, growth inhibition seemed to be triggered by interspecies differences between the parental genomes. These experiments raise the problem of the role of species-related genetic determinants which may interfere with the expression of certain phenotypes. It is an interesting aspect that was for the first time reported in this work, and has been overlooked in the interpretation of other experiments along these lines.

b. DNA Repair

The putative role of DNA repair in the mechanisms of cell aging and transformation has been extensively treated heuristically [Williams, 1976] and approached experimentally. The cultivation of cells in vitro has been a useful tool to test the hypothesis. Different aspects of the relationship between DNA repair and aging have been explored this way, including the possible implications for the division potential of cells. However, in spite of the considerable effort put into this line of research, the relationship between DNA repair processes and cellular aging are not yet understood. Actually, in many respects the results suggest that there is no relationship.

Different methodologies are available to evaluate the repair of DNA, which can reveal different mechanisms of repair. One of the methods widely used in cytogerontology was the ability to perform unscheduled DNA synthesis after UV irradiation. Fibroblasts have also been, in this regard, the working horse. Unscheduled DNA synthesis was mainly measured as a function of the number of grains found after autoradiography of cells labeled with ^{3}H-thymidine and treated with an inhibitor of scheduled DNA synthesis. This can create a pitfall, since hydroxyurea, the main inhibitor used, can also inhibit unscheduled DNA synthesis depending on the cell type [Collins et al., 1982] contrary to what was previously thought.

DNA Repair and Species Life Span – Intriguing results were obtained in regard to the life span of several mammalian species and the capacity of the respective fibroblasts to perform excision repair. Hart and Setlow [1974] analysed the cells obtained from skin biopsies of shrew, mouse, rat, hamster, cow, elephant and man. Both the initial rate and the maximum incorporation of ^{3}H-thymidine increased with the life-span of the species; unscheduled DNA synthesis was approximately proportional to the logarithm of life span (fig. 28, 29). Hart et al. [1979] extended this work to

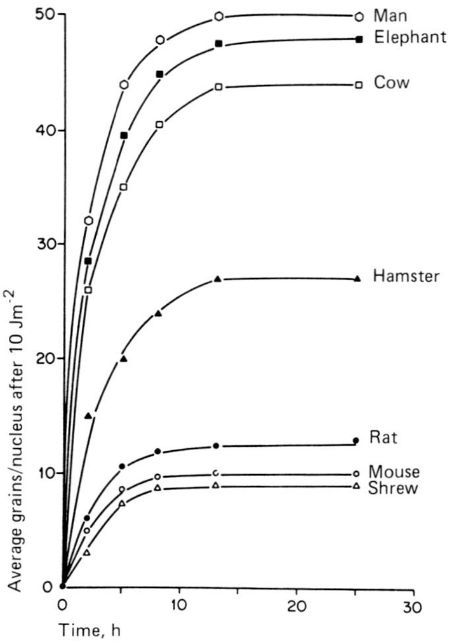

Fig. 28. Average amount of unscheduled synthesis as a function of time after 10 J/m^2 for seven different species [Hart and Setlow, 1974].

two rodent species of the same superfamily, which differ in life span by a factor of 2.5. The rate of DNA synthesis was 2.5 times greater in the species with a longer life span. The size of the repair regions (patch size), measured by bromouracil photolysis after UV irradiation, was the same for the two species, but the number of repaired sites was 2.2 times greater for the longer-lived species. Single-strand break repair after gamma-rays irradiation was not significantly different.

Kato et al. [1980] have reappraised the results of Hart and co-workers, on 34 species from 11 orders of mammals and found no correlation between repair potentials and life span. As pointed out by Kato et al., several reasons could explain the difference between their results and those of Hart et al. The fibroblasts used by Kato and co-workers originated from the lung, while those used by Hart came from the skin; moreover, no hydroxyurea was used in Kato's work and the animals were of different gerontological ages. Another difference that could be crucial is the use of cells from the second passage by Hart and Settlow, and after

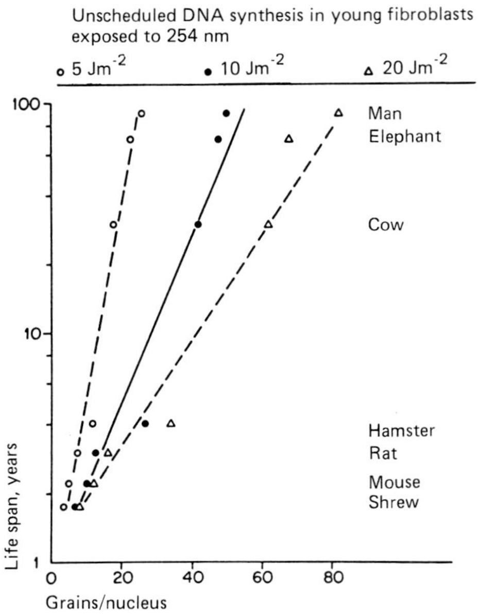

Fig. 29. Correlation between the amount of unscheduled DNA synthesis measured at 13 h after exposure to several UV fluences and the estimated life span of the species [Hart and Setlow, 1974].

several passages by Kato et al. Some works described below suggest that cells serially subcultivated in vitro may change in this regard.

Other investigators though, could confirm the results of Hart and co-workers. Francis et al. [1981] tested the cultivated fibroblasts from 21 mammalian species, mainly originating from the skin. In 2 cases, the kidney was the organ where the cells came from and in 1 case the liver. They found a good linear correlation between life span and the number of DNA excision repair sites, albeit several deviations from the linear relationship were observed; patch sizes were not related to maximal life spans. Treton and Courtois [1981] also found a good correlation between unscheduled DNA synthesis after UV irradiation of lens epithelial cells and the life span of rat, rabbit, dog, cow and horse. An identical correlation was found within a group of 6 different primates when not only primary fibroblast cultures but also lymphocytes were tested [Hall et al., 1984]. Maslansky and Williams [1985] analysed this problem in hepatocytes from five mammalian species and also found a positive correlation between this type of DNA repair and maximum life spans. At higher ultraviolet fluence levels, however, there were no differences in the amount of repair induced.

Hence, different groups could confirm Hart and Setlow's findings, although there might be exceptions for some species. No explanation, however, is yet available to elucidate this intriguing relationship. It has been speculated that the need for this type of repair evolved at a time when the planet was not protected by an ozone shield and only organisms capable of repairing the lesions induced by UV light, could develop [Burnet, 1978]. On the other hand, Walford [1979] suggested that the chromosomal locus controling DNA repair may be associated with other multigens important for longterm survival, such as those of the major histocompatibility complex; so that DNA repair, immune dysfunction and aging in higher animals, may be intimately related.

However, as discussed by Hart and Setlow [1974], unscheduled DNA synthesis after UV irradiation is just one form of repair and there must be much more to aging than just the failure of this mechanism. Furthermore, xeroderma pigmentosum (XP) patients who are defective in excision repair [Cleaver, 1969] do not seem to have premature aging. On the other hand, progeria patients show normal levels of excision repair.

Another type of repair, where interspecies differences have been described, is the removal of *Micrococcus luteus* UV-endonuclease-susceptible sites. Human fibroblasts are capable of this type of repair contrary to rat cells which almost completely lack it [Vijg et al., 1984]. No attempt has been made to relate this form of repair with life span. Vijg and co-workers, however, suggested that this characteristic may render the rat cells more dependent on postreplication repair systems which are error prone. It may lead to an increased frequency of genetic changes in rat cells, which could have a bearing with their susceptibility to transformation, described above. DNA repair abilities of mouse cells also differ from human [Yagi, 1982] and may, too, be implicated in their proneness to transformation [Elliot and Johnson, 1985].

DNA Repair and Aging in vivo. Cells cultivated in vitro have been used to check if DNA repair changes with the age of the donor and is deficient in diseases of premature aging.

Hennis et al. [1981] analysed unscheduled DNA synthesis in skin fibroblast cultures from normal human donors of different ages, measuring the relative capacity for repair of the three cell types found in these cell populations. The large pleomorphic and sail-shaped cells exhibited similar rates of unscheduled DNA synthesis (UDS), which was greater than that of the smaller spindle-shaped cells. This could be due to the increased DNA content of the former cells. No major changes in this type of DNA repair were observed for either type of cell with increased donor age.

Dell'Orco and Andersen [1981] compared fibroblasts from embryonic skin with those from the skin of a 73-year-old donor, also in regard to their ability to perform UDS. Cells from both donors maintained the ability to perform UDS throughout their life spans in vitro to the same degree, when tested in a confluent state of contact inhibition of growth. When arrested in a low serum concentration, postnatal cells displayed higher UDS. This work could be compatible with developmental changes that render the DNA more accessible to repair enzymes.

Human fibroblasts from donors of different ages were also compared in regard to survival of colony formation after UV irradiation and to the ability to reactivate UV-irradiated herpes simplex virus [Hall et al., 1982]. These results suggested that DNA repair mechanisms acting on lesions such as pyrimidine dimers, apurinic sites, alkylated bases and cross-links do not decline with aging. The authors pertinently pointed out that damage may accumulate any way during aging, in spite of the repair proficiency of mammalian cells.

No significant age-related differences were found either, in the rate and extent of the repair-replication response to UV radiation, of human epidermal keratinocytes [Liu et al., 1982]. In this study the 5-bromodeoxyuridine density-labeling method following UV irradiation, was used.

One group, however [Nette et al., 1984], found, using human epidermal cells, that the percent of cells capable of UV induced UDS does not decrease with donor age, but that the rate of UDS in individual cells decreases to a highly significant degree. The difference between this work and those mentioned above is that the latter authors used cells freshly separated from the skin, which attached to the culture vessel but did not proliferate in vitro. The other groups used cells maintained for at least a few doublings in culture. This could be of importance since Niedermüller et al. [1985] were able to find a distinct age dependency of excision repair in different rat tissues. In this work, excision repair was measured after damage by M-nitroso-methyl-urea.

A recent report described a decreased level of unscheduled DNA synthesis in lymphocytes and in fifth passage fibroblasts from a patient that was claimed to be affected with premature aging [Stefanini et al., 1986].

Treton and Courtois [1981] used rat lens epithelial in organ culture to evaluate the relationship between UV-induced UDS and development and aging. They found an age-related decrease in DNA repair, but most of it occurred up to 59 weeks of age and remained constant thereafter. In other words, it took place during the first third of the animal life span. It is interesting that the culture method used by Treton and Courtois is closer to that used by Nette et al. [1984], two works where the results seem closer to those obtained by Niedermüller et al. using tissues directly extracted from the animal.

UDS was analysed in primary cultures from mouse embryos of different gestational age [Peleg et al., 1976] and found to evolve with development. Cells from mice at 13–15 days of gestation excised thymine dimers as measured by paper chromatography and performed UDS after UV radiation. At late stages of development (17–19 days gestation), DNA repair capacity was impaired. These findings agree with those of Yagi [1982].

Fibroblasts from progeria patients have been extensively used to test the relationship between DNA repair and aging. These cells, however, seem to be efficient in all types of repair that were measured. Thus, Regan and Setlow [1974] found a normal ability to repair gamma-ray-induced single-strand breaks in DNA. The same observation was made by Bradley et al. [1976a] who could also not find any evidence in favor of DNA cross-linking with the alkaline elution method. Brown et al. [1980] used an immunofluorescence host cell gamma-irradiated virus reactivation assay, to evaluate the capacity of repair of fibroblasts from normal, progeria and XP donors. The latter cells showed the lowest level of repair; one progeria strain showed decreased reactivation and a second progeria strain showed normal reactivation. The results suggested a heterogeneity for DNA repair capacity among progeria fibroblast strains and that DNA repair is not a basic defect in this disease.

Normal levels of sensitivity to X-ray and UV killings and normal repair of X-ray induced single strand breaks and of UV damage, were also found in fibroblasts from Werner patients [Fujiwara et al., 1977].

Another progeroid syndrome (Rothmund Thomson) was also tested. Cultured fibroblasts from 4 patients displayed decreased survival following hypoxic gama-irradiation, measured as a function of colony-forming ability [Smith and Paterson, 1982]. The cells from 2 of the patients also displayed reduced sensitivity to the endonuclease of *M. luteus.*

Alzheimer's disease fibroblasts have also been studied; bleomycin-induced DNA double-strand breaks were efficiently repaired, but breaks resulting from treatment with N-methyl-N[1]-nitro-N-nitrosoguanidine were repaired more slowly.

DNA Repair and Aging in vitro. The role of DNA repair on aging in vitro, i.e. on the growth potential of cultivated cells, was also extensively analysed. It was first approached [Goldstein, 1971] checking the number of doublings in vitro of fibroblasts from XP, which are known to be defective in repair replication [Cleaver, 1968]. The life span of XP cells was found to be identical to that of age-matched controls. These results were confirmed be Cleaver [1984]. UV-light also failed to change the life span of normal cells in vitro, regardless of the age and culture stage of the cells at the time of exposure [Kantor et al., 1978; Dell'Orco et al., 1982].

Painter et al. [1973] analysed the capacity for UV-induced repair replication, of human embryonic lung fibroblasts at different population doubling levels. They concluded that a reduced level of DNA synthesis was found only at the last passage and thus, cannot be relevant for the mechanisms of aging in vitro. Old cells were competent for repair replication and strand rejoining after X-ray damage [Clarkson and Painter, 1974].

Hart and Setlow [1976] exposed the same cells to UV or gamma-rays radiation and to N-acetoxy-AAF, then UDS was measured. They came to the same conclusion as painter and co-workers. They also found a good correlation between cells that do scheduled and those that do unscheduled synthesis; which suggests that the failure to repair DNA of the terminal, phase IV cell, is related to the general failure of the DNA-synthesizing capacity. Identical findings were reported by Milo and Hart [1976] and Bowman et al. [1976a].

Mattern and Cerruti [1975] reported that isolated nuclei of WI-38 cells, 4–5 generations before the postmitotic state, i.e. in phase IV, lose the ability to excise damaged thymine residues from an exogenous gamma-irradiated or osmium-tetroxide-oxidised DNA substrate. Young and mid-passage cells removed damaged thymine to the same extent.

Pyrimidine dimer excision after exposure to UV radiation did not vary during serial population doublings of human embryonic lung fibroblasts [Cooke and Harris, 1981]. In this work, the removal of pyrimidine dimers was assessed by chromatography. The same authors did not find either, differences with population doubling levels, in the sedimentation velocity of newly synthesized DNA from cells pretreated with the chemical carcinogen 4-nitroquinoline-1-oxide [Cooke and Harris, 1982].

Still in the same type of cells (WI-38), the capacity to repair X-ray-induced DNA single-strand breaks was present even in terminally senescent cells [Bradley et al., 1976]. Suzuki et al. [1980], though found that the rejoining of single-strand breaks in terminal cells seemed to proceed more slowly.

Dell'Orco and Whittle [1978] measured UV-induced UDS through the life span in vitro of human postnatal skin fibroblasts kept in resting phase by lowering the serum concentration. They found that during the last two-thirds of the life span, arrested populations exhibited a 30–50% increase in the amount of detectable DNA repair. The authors postulated that the increase may be due to changes in DNA structure. It is interesting to compare these results with those reported above obtained by the same group, showing that under the same conditions postnatal cells of an old donor have a higher rate of repair than embryonic cells. It could mean that identical changes in DNA structure occur during serial proliferation in vitro and during aging in vivo. Dell'Orco and Whittle [1981] obtained further evidence that the increased repair of DNA in low-serum arrested cells is due to a different conformation of DNA. Indeed, they found that there was no age-related difference in the number of UV-induced *Micrococcus* endonuclease-sensitive sites but the removal of these sites was more rapid in late passage cells. High population doubling level nonirradiated control cells, treated with endonuclease, also had a smaller molecular weight, which shows that after serial divisions there is an increased number of endonuclease sensitive sites. The authors suggested that the elevated level of UDS may result from the excision not only of UV-induced damage, but also of other altered DNA sites. These results show how other factors such as DNA structure can influence the evaluation of UV-induced damage and again point to the idea that a fundamental change occurring in DNA is related to its conformation.

The evolution of DNA repair during serial doublings in vitro was also evaluated in cells from species other than the human. As mentioned above, other factors have to be taken into account in the case of rodent cells which have a high probability of escaping senescence.

Meek et al. [1980] followed mouse fibroblasts cultures through the growth decline and growth recovery up to the time when the cells immortalized. They found that the capacity to perform UV-induced UDS followed the same pattern as scheduled DNA synthesis, i.e. it first declined with passage number and then increased in cultures which spontaneously became continuously proliferating cell lines. A similar observation has been reported by Little [1976] for hamster cells. So rodent cells seem to behave differently from human cells where UDS declines only at the very end. In rodent fibroblasts which age more rapidly in vitro, the heterogeneity of the cell population in regard to the capacity to synthesize DNA is more pronounced, and postmitotic cells may become apparent earlier.

La Belle and Linn [1984] also found that the response of mouse cells to UV and N-methyl-N^1-nitro-N-nitrosoguanidine decreased rapidly within the first two doublings and could not be detected at the 12th doubling. Uracil DNA glycosylase activity also declined during growth crisis. AP endonuclease activity did not change up to the 12th doubling.

Vijg et al. [1986] followed UV-induced DNA excision repair in rat fibroblasts, determined autoradiographically as UDS and by means of the BrdU photolysis assay for the evaluation of the number and size of repair patches. They found first an increased UDS associated with a morphologic shift from small spindle-shaped cells to large pleiomorphic ones with larger nuclei. These results are reminiscent of those mentioned above, obtained by Hennis et al. [1981], which were attributed to an increased DNA content of the pleiomorphic cells. In the work of Vijg et al. [1986], however, the number of repair patches, which is expressed per amount of DNA, was also substantially higher in the secondary cells. In cell populations in growth crisis, UDS was found to decline, and after immortalization it reached levels identical to those of the original primary cultures.

Rat fibroblasts have also been used to analyse DNA repair in cells treated with 4-nitroquinoline-1-oxide, a chemical which mimics UV light in several of its biological effects [Chan and Walker, 1977]. Repair synthesis was measured by the incorporation of ^{3}H-thymidine into DNA isolated by isopicnic centrifugation in CsCl; repair of single-strand breaks was evaluated from the sedimentation velocity of DNA in alkaline sucrose gradients. By the seventh subculture, repair synthesis was about 15% of its starting value, while repair of single strand breaks was not detectable.

Hence, besides the interesting relationship between one form of DNA repair and the maximum life span of different species, the results are rather meager in regard to the mechanisms of aging and those responsible for the growth potential of somatic cells. As pointed out by Cleaver

[1978] DNA repair depends to a large extent upon DNA synthesis and the decline in repair in the postmitotic cell must be secondary to the defects that alter DNA replication. It confirms the results, obtained by ourselves and reported in the chapter describing the kinetics of proliferation, that the loss of the DNA synthetic activity is not exponential and occurs only at the very end. Non-DNA-synthesizing cells that are produced at each population doubling [Rao, 1981] must represent a very low fraction and be mostly lost.

The results also brought up evidence favoring the idea that conformational changes in DNA become apparent during development in vivo and serial proliferation in vitro [Dell'Orco and Whittle, 1981]. Some experiments raise the possibility that cells that proliferate in vitro might reacquire some forms of repair that decline with development and/or aging in vivo.

Finally, there are obvious interspecies differences in DNA repair ability. It would be interesting to know if they are responsible for some of the different biological properties described in the preceding chapter, in particular for the variable susceptibility to transformation and probability of spontaneous immortalization.

c. Protein Metabolism

Testing of the Error Hypothesis. The idea that cellular aging could be due to an accumulation of errors during protein synthesis has been one of the most popular theories, if not the most popular. The appeal of this theory is probably due to the deeply rooted feeling of our civilization that human kind fails and degenerates because of errors commited. In fact, in some instances the desire to prove the theory was such that any change in a protein was interpreted as evidence in favor of errors, when many other hypothesis could explain the data. So far the search for errors in proteins has been elusive and all the experiments claiming to support their presence were of an indirect nature.

The evaluation of the thermolability of enzymes was used by Holliday and Tarrant [1972] who reported an increased thermolability of glucose-6-phosphate dehydrogenase (G6PD) and 6-phosphogluconate dehydrogenase at higher population doubling levels of human embryonic lung fibroblasts. Moreover, the RNA base analogue 5-fluorouracyl accelerated senescence of this cell population, which was preceded by the appearance of enzyme with increased thermolability. These results were thaught to support the error theory. The same group [Holliday et al.,

1974] found an increased percentage of heat-labile enzyme in skin fibroblasts from patients with the Werner's premature aging syndrome. A finding which was confirmed by Goldstein and Singal [1974a] for G6PD who also reported a decreased expression of two histocompatibility antigens, HL-A2 and HL-A9. It was concluded that these 3 different gene products are defective in this disease. Later, Goldstein and Moerman [1975] extended their findings to 6-phosphogluconate dehydrogenase and hypoxanthine-guanine-phosphoribosyltransferase. They also reported an increase in heat-labile enzymes in skin fibroblasts from subjects with progeria. These results could not be confirmed both in erythrocytes and in the fibroblasts from these patients and from Werner's syndrome patients [Brown and Darlington, 1980, and personal commun.]. A normal thermostability of hypoxanthine-guanine-phosphoribosyltransferase has also been found in erythrocytes from Werner's syndrome patients [Takeuchi et al., 1986].

Other investigators considered other possibilities for the increased heat-lability of proteins and looked for possible mechanisms. Kahn et al. [1977] could ascertain that G6PD from young and senescent cells had the same heat lability after purification and that dialysis of the crude extracts from senescent cells improved the heat stability. Heat lability could be induced on the purified enzyme by the crude extract from senescent cells deprived of the enzyme, and by protein fractions purified from senescent cells. Obviously, this elegant series of experiments showed that the enzyme alteration must be due to postsynthetic events created by the intracellular environment rather than errors due to misincorporation of aminoacids.

Duncan et al. [1977] studied the different forms of this enzyme separated by acrylamide gel electrophoresis and found two bands, one corresponding to the dimeric form and the other to the tetrameric form. The latter, which was more heat sensitive, increased as the cells became senescent in the same proportion to the fraction of heat labile enzyme. Hence, the enzyme alteration was attributed to a shift in the equilibrium between the two forms.

The same investigators [Duncan and Dell'Orco, 1979] studied the effects of the amino acid analogue *p*-fluorophenyl-alanine. In the presence of the analogue there was a rapid accumulation of heat-labile G6PD molecules to levels observed in senescent cells but without a shift in equilibrium between the molecular forms of the enzyme. Treated cultures achieved as many population doublings as untreated controls.

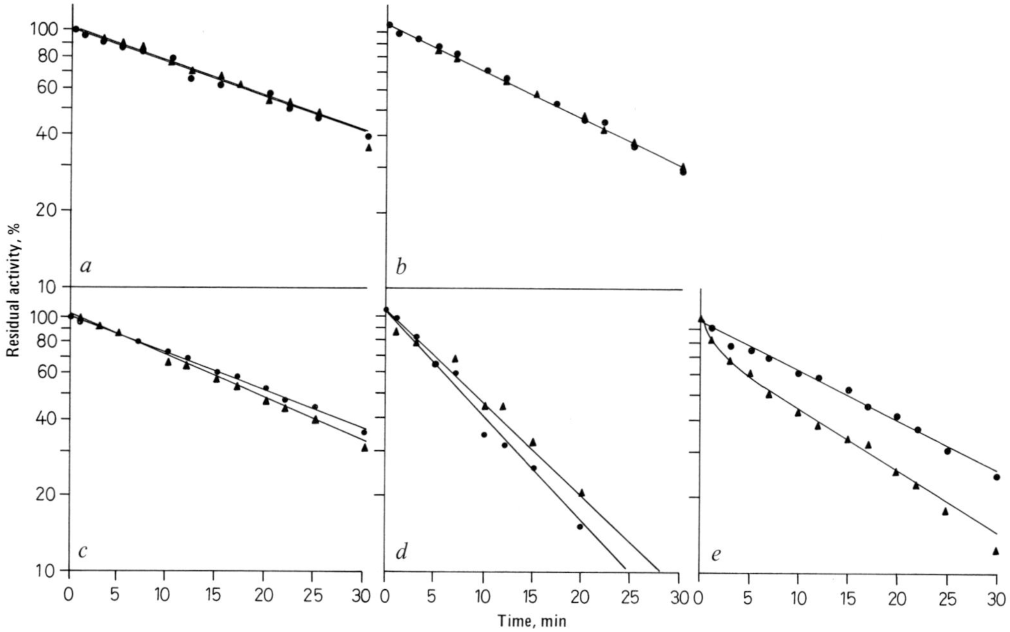

Fig. 30. Inactivation curves of enzymes from young and old human embryonic lung fibroblasts. ● = Cells from passages 22–25; ▲ = cells from passages 42–45. *(a)* N-acetyl-alpha-*D*-galactosamine; *(b)* N-acetyl-beta-glucosamine, and *(c)* alpha-*D*-glucosidase are lysosomal enzymes; *(d)* sulphite cytochrome c reductase, and *(e)* glucose-6-phosphate dehydrogenase are located in the mitochondria and cytoplasm, respectively [Houben and Remacle, 1978).

Remacle and co-workers made a thorough investigation of the mechanisms involved in enzyme heat lability. They first reported [Houben and Remacle, 1978] that only cytoplasmic enzymes displayed heat lability, lysosomal and mitochondrial enzymes did not present this alteration (fig. 30). This was so in the case of human embryonic lung fibroblasts aged in vitro and also for skin fibroblasts from Werner's syndrome patients [Houben et al., 1980]. Thus, these data suggested that some cytoplasmic factor was responsible for the alteration and buttressed the conclusion of Kahn et al. [1977].

Further support for cytoplasmic factors was obtained by the same group [Houben et al., 1984a] from the observation that incubation at 4 °C and pH 7.4 induced the appearance of heat lability, and that an acid pH and the presence of NADP or glutathione protected the enzyme from being modified. They also detected with electrophoresis the dimeric and

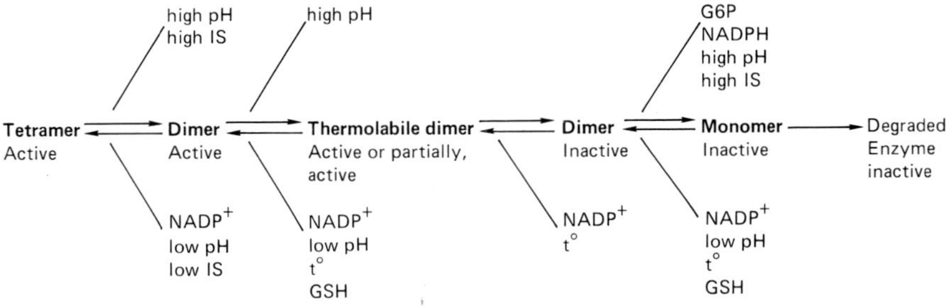

Fig. 31. Model for the modifications of glucose-6-phosphate dehydrogenase during aging of human embryonic lung fibroblasts in culture. IS = Ionic strength [Funfu et al., 1985].

tetrameric forms. The study of enzyme alteration, inactivation and its reversibility showed that inactivation was linked to the dissociation into inactive monomers whereas during reactivation the monomers reassociated into dimers and tetramers [Houben et al., 1984b; Funfu et al., 1985]. According to their results the altered form of the enzyme is dimeric. Cytoplasmic superoxide dismutase (SOD) but not mitochondrial SOD displayed also increased heat lability which in this case was associated with the tetrameric form [Somville et al., 1985].

This shift in the different forms of the enzyme could correspond to the normal pathway of protein degradation [Rothstein, 1977] (fig. 31). Indeed, it is now known that some proteins can assume a functional heterodimeric form or a homooligomeric one targeted for degradation [Woods and Lazarides, 1986]. But other mechanisms of protein degradation are also known [Bachmair et al., 1986].

The works of Tollefsbol et al. [1982] and Chapman et al. [1983] also favored the interpretation that the enzyme alteration is due to the accumulation of forms of the protein that are part of the degradative pathway. At least one glycolytic enzyme, triose-phosphate isomerase accumulates in old cells as an unstable form due to specific deamidation of two asparagines, another mechanism available for protein degradation [Yuan et al., 1981].

DNA polymerases alpha-a and alpha-b were also found to become heat labile in aging cultured chick embryo fibroblasts [Fry and Weisman-Shomer, 1976]. This change, however, was not accompanied by modification of their specific activity, chromatographic behavior or their abilities to utilize deoxyribonucleoside triphosphates with activated DNA template. Additional changes concerned a decreased ability of DNA polymerase alpha-a to copy oligo (dG)-poly(dC) while that of polymerase alpha-b increased. Three hypotheses were proposed to explain the modifications observed with the two polymerases: (1) changes in the amount of nonpolymerase contaminants, (2) postranslational modifications such as protein cleavage, phosphorylation, carbamylation or acetylation, and (3) presence of isoenzymes produced by genes different from those coding for the DNA polymerases of young cells.

Some investigators, however, could not find altered enzyme molecules during in vitro aging. Thus, Danot and Gershon [1975] reported that the amount of aldolase per cell as well as the amount of enzyme per unit of antigen remained constant throughout the life span of mouse embryo fibroblasts. The aldolase from all population doubling levels was also equally sensitive to heating at 60 °C. Hence they could not confirm the claim of Lewis and Tarrant [1972] for human embryonic fibroblasts, of a progressive loss in specific activity of lactic dehydrogenase (LDH), not associated with a loss of the enzyme protein. The quantity of enzyme was expressed in terms of units of antigenic activity (cross-reacting material) and of units of catalytic activity. LDH is not the appropriate enzyme for this type of studies because of the presence of different isoenzymes. Actually, a wide variation in LDH activity has been found during senescence of human fibroblasts [Minkowitz et al., 1985] with a final increase which could be due to variations in the different forms of the enzyme. Lewis and Tarrant had also found a loss with cellular senescence of the discrimination between an amino acid and its analogue for incorporation into proteins. They claimed that their results supported albeit did not prove the error catastrophe theory. On the other hand, Ryan et al. [1974] grew cells continuously in the presence of nontoxic doses of the amino acid analogues *p*-flurorphenylalanine and ethionine, to find that the cultures completed the same number of population doublings as the controls.

Pendergrass et al. [1976] also could not observe an increase in enzyme thermolability in postnatal skin fibroblasts aged in vitro; although an increased thermolability could be detected for the G6PD from a thermolabile mutant and for the G6PD from cells treated with a proline analogue. Titration of G6PD activity with neutralizing antibody also failed to detect an increment of enzymatically defective enzyme in old cells. Identical results were reported for the enzyme phosphoglucose isomerase by Shakespeare and Buchanan [1978].

Thus, it seems that the appearance of an increased enzyme thermolability might depend on the procedures used to maintain the cells. This possibility is also suggested from the results of Somville et al. [1985] who could decrease the thermolabile fraction of cytoplasmic SOD, through the stimulation of cell division by vincamine.

Another result which further questions the usefulness of using enzyme thermolability to identify changes in a protein is that obtained by Viceps-Madore and Cristofalo [1978]. They observed that although there is a gradual decline in glutamine synthetase specific activity, the enzyme from younger cells is more heat labile than that from older cultures.

Sambuy and Bittles [1984] examined six exopeptidases during the life span in vitro of a human embryonic lung fibroblast cell line. The biphasic nature of the heat-inactivation curves suggested the presence of more than one isoenzyme. At the end of the incubation time the residual activity was lowest in late passage cells. However, the individual exopeptidases did not show any change in mobility on polyacrylamide electrophoresis nor in their relative activity and substrate specificity.

Goldstein and Csullog [1977] pulse labeled at 42 and 37 °C postnatal human skin fibroblasts with ^{3}H-thymidine, ^{3}H-uridine and ^{14}C-phenylalanine and could not find an increased heat lability in the profiles of precursor incorporation. If the enzymes implicated in these metabolic pathways were modified, it should have been possible to detect variations in the incorporation of the precursors at different temperatures.

Virus replication in young and old cells has also been used as a probe for the detection of protein synthetic errors. Holland et al. [1973] used vesicular stomatitis and herpes simplex viruses and found that the virus yield did not vary even in terminal cells. The thermal inactivation of the infectivity was identical in virions produced in young and old cells. The spontaneous rate of mutation to resistance to the inhibitory effect of guanidine on virus replication was also the same. Tomkins et al. [1974] reported that with poliovirus and herpesvirus the amount of virus produced, pattern of cytopathology and plaque morphology did not vary with cell senescence. The herpes virus proteins analysed by polyacrylamide gel electrophoresis were also identical. The same results were obtained by Pitha et al. [1975] on human embryonic fibroblasts in vitro and human skin fibroblast cultures from donors of different ages [Danner et al., 1978]. Hence, with this methodology support for the protein error theory could not be obtained either.

The fidelity of protein synthesis was also evaluated from the translation of polyuridylic acid by cell-free extracts from human embryonic and postnatal fibroblasts [Buchanan et al., 1980; Wojtyk and Goldstein, 1980]. Since the UUU codon specifies phenylalanine, faithful translation of poly U leads to incorporation of phenylalanine alone into polyphenylalanine [Nirenberg and Matthis, 1961]. If errors occur, leucine, for instance, could become incorporated since its codons UUA, UUG and CUU have a degree of similarity to the UUU codon. No difference could be detected in the misincorporation of leucine with aging in vitro. The antibiotic paromomycin had a profound effect on the leucine misincorporation, but the latter was of the same degree during the whole cell population life span [Buchanan et al., 1980]. No difference could be detected either in the translational fidelity during in vitro aging of postnatal cells from normal donors and in cells from progeria and Werner's syndrome patients [Wojtyk and Goldstein, 1980]. The same type of experiments were performed on a population of skin fibroblasts and on clones isolated from the mass culture [Wojtyk and Goldstein, 1982]. A decline in error frequency was observed in the mass culture; in 12 individual clones, however, error frequencies did not decline and in two clones they increased. The error frequencies showed no correlation with the maximum doubling number but there was a weak correlation with growth vigor. The authors suggested that selective pressures may favor more vigorously dividing clones with low protein synthetic errors. Still, it is difficult to understand the opposite results obtained with the mass culture and the cloned population.

Misincorporation of ^{35}S-methionine into histone H1 was also used to evaluate the fidelity of the synthesis of this class of proteins, since it is supposed not to contain any methionine or cysteine [Buchanan and Stevens, 1978]. A methionine-containing fraction associated with H1 increased with age of the culture. It was thought to be due either to the loss of activity of the enzyme that hydrolyses the N-terminal methionine, or represent a new nuclear protein present in nondividing cells. Medvedev et al. [1977] have also reported an increase with in vivo age of the methionine-containing subfraction of histone H1 from rat liver and spleen.

Putative protein synthetic errors were also assayed on two-dimensional gels after induction of mistranslation by amino acid starvation [Harley et al., 1980]; the rationale for this methodology being the change in isoelectric points that should result from amino acid substitutions. Similar or lower error frequencies were found in late passage cells from fetal, young and old normal donors. Cells from subjects with progeria or Werner's syndrome did not have elevated error frequencies either. Although these experiments concerned only actin protein synthesis, the authors concluded that they ruled out a general increase in the error frequency of protein synthesis.

The fidelity of DNA polymerase activity was also approached experimentally using a nonphysiological divalent cation, i.e. manganese [Linn et al., 1976] or a physiological one, i.e. magnesium [Murray and Holliday, 1981]. The data obtained in both works suggested that cells aged in vitro are more error prone. In these experiments, however, the method used for isolating DNA polymerases yields a lower molecular weight enzyme, probably produced by proteolytic degradation and resulting in a conformational change of the protein [Murray, 1981]. This enzyme has an increased in vitro error frequency compared to the higher molecular form obtained by an extraction procedure which prevents proteolytic degradation [Murray, 1981]. The proportion of the small molecular weight enzyme rises from 7% in young cells to 29% in cells aged in vitro.

Additional evidence in favor of an unaltered protein synthetic machinery was provided by the observation that terminal human embryonic lung fibroblast synthesize heat shock proteins as well as young cells [Tsuji et al. 1986].

Hence, it seems that for the moment, the evidence against the idea that cellular senescence is due to increased errors in the synthesis of proteins is overwhelming.

Rate of Protein Synthesis. One of the constant findings of cell senescence is an increased volume and protein content both with in vitro and in vivo aging [Segal and McCoy, 1974; Kaftory et al., 1978; Shevitz et al., 1986]. Why this takes place, though, is completely unknown. Since protein content increases by a factor of 1.8 and cell volume by a factor of 4.7, it means that the protein concentration is considerably diluted [Kaftory et al., 1978]. The dilution of cellular constituents must create new topological constraints, and this way be responsible, to a large extent, for the manifestations of cellular senescence. Indeed, such spatial modifications must induce significant variations in the association between molecules and be correlated with conformational modifications at the molecular and supramolecular level.

The amount of proteins per cell depends upon their rate of synthesis and degradation and their transport and excretion across the cell membrane. The rate of synthesis is influenced inter alia by the free intracellular amino acid pool and by RNA synthesis. Significant in vitro age-related decreases were found in the pool of five essential amino acids (tyrosine, phenylalanine, leucine, isoleucine and valine) and an increase of methionine [Sambuy and Bittles, 1982]. According to these investigators, the changes in amino acid pools could be due either to a disturbance of the operation of the amino acid transport systems or to the availability of tRNA species or tRNA synthetases. Concerning the synthesis of total RNA the results are conflicting and a decrease [Macieira-Coelho et al., 1966a; Srivastava, 1973; Razin et al., 1977; Evans et al., 1978; Nose and Okamoto, 1980], an increase [Schneider et al., 1975] and no variation [Hill, 1976] have been reported. Some disturbance, though, must take place since less ribosomal RNA is synthesized per cell [Macieira-Coelho and Loria, 1974, 1976; Bowman et al., 1976b]. The increased ratio of 4S transfer RNA to 18S ribosomal RNA [Johnson et al., 1977] also suggests that RNA synthesis is disturbed. Total RNA per cell, per protein and per DNA increases [Cristofalo and Kritchevsky, 1969; Johnson et al., 1977]. An increased RNA turnover and wastage of ribosome precursors have also been described [Wolf et al., 1980; Michl and Svoboda, 1967]. On the other hand, the stability of some messenger RNAs seems to be decreased with in vitro aging [Stein and Burtner, 1975]. This was concluded from experiments where young and old human embryonic fibroblasts were stimulated to divide in the presence of actinomycin D. An increase in nonhistone chromosomal protein synthesis, an early event following stimulation, took place in young but not in old cells. This suggests that only in early passage culture are these proteins synthesized on pre-existing templates.

Another event that influences protein synthesis is the rate of elongation of the peptide chain. Both the amount and the catalytic activity of the elongation factor EF-1 alpha, the crucial enzyme in this step, decrease late during the life span of human embryonic fibroblasts [Cavallius et al., 1986]. The synthesis of EF-1 alpha, however, does not show age-related variations. Since the decline in amount and catalytic activity also occurs in low-serum arrested young cells, it could be due to the nondividing state rather than be characteristic of senescence. This is suggested from the further reduction in enzyme activity in old cells maintained in low-serum medium, a reduction which is reversible. This further reduction in old cells could be due to the effect of low-serum concentration on residual cells with short cycles known to be present in these cultures (see chapter IV).

The rate of overall protein synthesis has been studied in cultivated fibroblasts both with in vitro and in vivo aging. Amino acid incorporation into acid precipitable material was found to decrease progressively during serial proliferation of chicken embryonic fibroblasts expressed on a per cell basis [Macieira-Coelho and Lima, 1973; Kaftory et al., 1978]. The latter authors compared sparsed cells, so that the result could be due to the reduction of the dividing compartment in the senescent cells. But Macieira-Coelho and Lima measured cells at different population doubling levels from the time of inoculation to resting phase and obtained identical results. This decline in amino acid incorporation into acid-precipitable material occurred earlier than the decline in RNA synthesis [Macieira-Coelho and Lima, 1973].

In human postnatal skin fibroblasts, however, a progressive increase in total protein synthesis per cell has been reported, without a change in leucine incorporation/mg protein [Shevitz et al., 1986]. After the last population doubling the increase in protein synthesis per cell was 4-fold.

It would not be surprising if these conflicting results would be due to interspecies differences since chicken and human fibroblasts differ in many other respects. However, other authors reported a decreased protein synthesis in human embryonic cells aged in vitro [Ballard and Read, 1985]; so that the differences could be between postnatal versus embryonic cells.

Human fibroblasts from donors of various ages were also compared in this respect but the data were erratic [Chen et al., 1980]. There was a higher incorporation in cells from fetal, newborn, 10-month-old and 4-year-old donors. Postnatal cells from 5 donors ranging from 10 to 90 years old showed a lower leucine incorporation but cells from a 35-year-old donor displayed higher incorporation. Hence, nothing can be concluded from this work and a higher number of samples would be required. Leucine incorporation has also been found to decrease to 67.5% of the control level in early passage cells from aged donors as compared to those from young; in fibroblasts from Alzheimer's patients the decline was 88.9% [Peterson and Goldman, 1986]. But in these experiments sparse cultures were compared so that the differences may be due to the respective sizes of the cycling and noncycling cell compartments.

It is surprising that even such a simple feature as the measurement of protein synthesis can yield opposite results. This is representative of the difficulties in the standardization of tissue culture methods and of the differences between apparently identical cell types. In our experience, the maintenance of standard conditions is a daily task in a tissue culture laboratory that requires a close watch on all parameters, from the water used to prepare the nutrient medium to the flasks where the cells grow which are sometimes changed by the manufacturers without further notice to the consumer. In addition, extreme care should be taken in using cells for the experiments; cultivated cells display sometimes variations whose cause remains unknown. If care is taken in using cells that display normal characteristics, results are perfectly reproducible. Latent contaminations, as for instance with mycoplasma, are also frequently overlooked.

As will be described in another chapter, differences between cell types morphologically identical should also be considered. Fibroblasts, for instance, have different properties depending on the body site, a feature that could explain many opposite data. Cultures should also be used under identical physiologic conditions in order to minimize parameters other than senescence that may influence the results. The growth stage of the cultures for instance is a source of disparity. As will be described in the following chapters, dividing cell populations in culture, as in the organism, are in permanent evolution; their properties and heterogeneity change continuously and two cultures at the 10th and 40th population doubling, would not be comparable if used for the experiments on the second day after seeding.

The synthesis of specific proteins has also been analysed. It was reported that the dependency upon ascorbate for the production and hydroxylation of collagen is influenced by serial divisions of cultivated cells [Houck et al., 1971; Paz and Gallop, 1975]. Ascorbic acid reduces to half the amount of collagen produced by young cultures and increases 2-fold the hydroxylation of this large structural protein. In older cultures, though, the amount produced is unaffected by the presence of the vitamin but the cells become more dependent on its presence for hydroxylation. In terminal cultures hydroxylation is low regardless of the presence of ascorbic acid [Paz and Gallop, 1975]. These experiments show very clearly that it is the regulation of the synthesis of this protein and its processing that are modified during senescence.

Terminal postnatal skin fibroblasts and cells from Werner's syndrome patients also produce a collagen with a subunit molecular weight larger than that of procollagens [Basler et al., 1979]. The authors suggested that this protein could be a form of procollagen produced by incorrect maturation of mRNA or incomplete processing of procollagen.

The collagen synthetic activity has been reported to decrease in human fibroblasts [Houck et al., 1971; McCoy et al., 1982] or to increase [Breul et al., 1980] when evaluated from the incorporation of proline into collagen and noncollagen protein. Mollenhaur and Bayreuther [1986] also found an increased production of collagen with aging in rat fibroblasts. Hildebran et al. [1983] studied the synthesis of collagen with methodologies supposed to avoid precursor pool problems. They analysed tRNA proline and the free proline pool specific radioactivity, and total incorporation of proline into cellular protein and collagen. The overall protein synthesis was slightly lower in cells aged in vitro but was not statistically significant. There was, however, a significant reduction in collagen synthesis. The collagen content, though, was increased in aged fibroblasts and the authors concluded that there must be an associated reduction in the rate of degradation. Total protein and collagen synthesis was analysed by Johnson et al. [1986] in gingiva fibroblasts from donors ranging from 12 to 68 years. Protein and collagen production decreased more than 5-fold in confluent, resting cultures, as a function of donor's age. The pattern of protein produced was unaffected by donor's age, as judged from two-dimensional electrophoresis. The slowdown in protein synthesis was not selective, it seemed to affect all proteins produced.

The synthesis of fibronectin, on the other hand, was reported to be increased per cell, during aging in vitro of postnatal human skin fibroblasts [Shevitz et al., 1986].

A developmental rather than age-related change in collagen synthesis has also been reported. Indeed, Sage [1985] found a unique low molecular weight collagen in fibroblasts from fetal bovine skin. It was preferentially synthesized by cells isolated from first trimester embryos at early passage. At later passages these cells behaved in this respect like those isolated from late trimester embryos. Hence, this is a developmental change taking place both in the embryo and when the embryonic cells are serially proliferating in vitro.

Protein synthesis has also been analysed during aging of cultivated lens cells [Ramaekers et al., 1984]. The relative amount of vimentin decreased as compared to actin during the elongation process that characterizes senescence of these cells, an event that also occurs in vivo. An increased synthesis of a 26-kdalton protein was observed together with a decrease of a 33-kdalton band seen on SDS-gel electrophoresis. Overall glycosylation of proteins decreased.

Protein Degradation. Growth factors are potent stimulators of protein synthesis, and inhibitors of protein degradation. During in vitro senescence of human embryonic fibroblasts, protein synthesis becomes much less responsive to any growth factors or to fetal bovine serum, but the inhibitory response on protein degradation is retained [Ballard and Read, 1985]. These investigators suggested that the reduced effect of growth factors on DNA synthesis in ageing cells could be a consequence of the inability to achieve a rate of protein synthesis that exceeds protein breakdown.

Protein degradation has also been extensively studied and unfortunately the data obtained by the different groups do not always agree.

Bradley et al. [1975], following their finding of an increased proteolytic susceptibility in WI-38 cells after the last doubling, studied protein degradation in the same cells in resting phase in the absence and in the presence of amino acid analogues [Bradley et al., 1976b]. Towards the end of the cell population life span, short-lived proteins were degraded more rapidly than long-lived ones. After the last doubling both type of proteins were degraded faster. Analogue-containing proteins were degraded faster but old cells did not seem so efficient in their degradation. Essentially, the same observations were made by Shakespeare and Buchanan [1976]. They also found that the increased proteolytic activity was associated with a lysosomal fraction [Shakespeare and Buchanan, 1979].

As pointed out by Dean and Riley [1978] in several of these experiments the cold analogue is used together with a radioactive amino acid so that the turnover of one precursor is not separated from that of the other. They avoided this problem using a radioactive analogue (canavenine) and found that proteins containing the arginine analogue were degraded faster in old cells. No variation was observed during the human embryonic lung fibroblasts life span, in the ability to degrade either normal or analogue-containing proteins. There was, however, an accelerated degradation, with senescence, of normal short half-life proteins. In this work, though, sparse cells were compared at different population doubling levels 48 h after seeding, a situation where the dividing fraction varies significantly.

Goldstein et al. [1976] measured proteolysis in confluent cultures after a medium change. This stimulates the cultures to a different degree depending on their age so that the results depend on the protein degradation occurring in the G1 of nondividing cells and in the G1 of those induced to divide. Thus, they cannot express modifications taking place during senescence.

Dell'Orco and Guthrie [1976] used a better approach by placing confluent cultures in a nonmitotic state induced by a nutrient medium containing 0.5% serum. During the first 4 days there was no age-related difference in the synthesis and turnover of proteins. After day 4, late passage newborn forskin fibroblasts, maintained a constant protein content but with faster rates of both protein synthesis and turnover. This occurred at the midpoint of the in vitro life span.

Kaftory et al. [1978] approached the problem in chick embryo fibroblasts cultures and found a slowdown of the average rate of protein degradation. The breakdown of abnormal puromycin peptides proceeded at the same rate in young and old cultures. They compared, though, cultures in sparse conditions.

In order to avoid problems related to the reutilization of radioactive amino acids and to putative different proteins in young and old cells, Dice [1982] evaluated degradation of microinjected proteins. He found that in vitro aged human embryonic fibroblasts, degraded more slowly RNAases A and B, aldolase, lysosyme and the synthetic polyglutamate:tyrosine:alanine. Confluent resting phase cultures were used in these experiments. This is an elegant approach and the one that is less influenced by artefacts if one excludes the trauma of the microinjection. The results, though, do not give any information about the degradation of altered proteins. It was suggested that the increased lifetime of the protein may increase the probability of being post-translationally modified. The same investigator reported that ginseng extracts inhibit intracellular degradation of long-lived proteins but not of short-lived ones, an effect identical to that they obtained with growth factors [Lu and Dice, 1985].

The decreased ability of dyalyzed serum to suppress protein degradation in aged cells is also suggestive of alterations of this aspect of protein metabolism [Wharton and Riley, 1986].

Transport. Finally, the transport of proteins across the cell membrane was also studied. The membrane was found to become more permeable to bovine serum albumin [Berumen and Macieira-Coelho, 1977]. Stimulation of albumin uptake by polyornithine was also increased, an event that took place simultaneously with increased protein leakage. It was suggested that protein intake takes place in parallel with excretion of other proteins, that this pump mechanism is altered during senescence, and that it contributes to the increase protein content and cell volume. The changes in protein transport, however, must be selective since it is not affected in regard to insulin [Press and Pitha, 1974].

In conclusion, one can say that the data does not favor the concept that cell functions are modified during senescence, due to the accumulation of protein synthetic errors. It seems rather that the regulation of their synthesis, processing and degradation is altered, but the exact disturbance remains to be elucidated. These modifications do not take place with the same pattern for all proteins. The transport of proteins into the cell and their excretion is also affected in a selective fashion. Senescence could be the result, inter alia, of a discoordination of these different aspects of protein metabolism.

The experiments attempting to detect the synthesis of new proteins, are reported in the chapter describing the terminal stage (phase IV).

d. Free Radicals

The idea that the accumulation of molecules modified by free radicals are one of the main causes of cellular aging was put forward by Harman in 1956. The cell culture system has also been used to test the theory but nothing relevant either to support or invalidate the concept has come up so far.

One of the associations between in vitro and in vivo aging in regard to free radicals comes from the accumulation of lipofuscin in lysosomes in both situations.

One of the effects of free radicals is the peroxidation of lipids which gives rise to aldehydes. Aldehyde groups can form Schiff bases and cross-link different molecules with the production of large polymers that are not digested in the lysosomes. These molecules have a characteristic fluorescence when excited in the 360–380 nm region giving emissions in the 440–470 nm range [Fletcher et al., 1973]. The accumulation of these brownish pigment is directly related to the age of the animal [Tappel, 1973]. Acid phosphatase has also been found in the lipofuscin granules of aging tissues [Gedigk and Bontke, 1956] and to be increased in the livers of aging animals [Zorzoli, 1955] suggesting an association between lipofuscin and lysosomes. These findings supported the idea that free radicals are implicated in aging in vivo.

The problem in vitro was approached by Cristofalo et al. [1967] measuring acid phosphatase activity in human lung fibroblasts; higher levels were found in cultures derived from adult lung than in those derived from fetal lung. The enzyme activity also increased during in vitro aging. Cristofalo and Kabakjian [1975] described later that the enzyme activity was increased not only in the lysosomal fraction, but also in the nuclear and microsomal fractions and in the supernatant, which suggested a decrease in enzyme latency.

Moreover, ultrastructural studies showed that in human fibroblasts aged in vitro, there is an accumulation of lysosomes and residual bodies [Brock and Hay, 1971; Lipetz and Cristofalo, 1972; Brandes et al., 1972; Robbins et al., 1970]. They can be observed, however, only in cells in the terminal stage and not in mid-passage cells [Boak et al., 1983].

Macieira-Coelho et al. [1971] found that acid phosphatase also increased in young cultures kept for long periods in resting phase, without renewal of the nutrient medium. Simultaneously, lysosomal-like structures seen with electron microscopy also increased. Upon stimulation of the cells, with a renewal of the nutrient medium, both the acid phosphatase levels and the lysosomal structures declined. Macieira-Coelho and Lima [1973] also measured acid phosphatase activity during aging of chick fibroblasts in vitro and found an increased enzyme activity only at the very end, when the cells stopped dividing (fig. 32). Enzyme latency did not vary. They concluded that the increase in acid phosphatase may be related more to the nondividing state rather than to the aging process.

Brunk et al. [1973] reached identical conclusions studying aging in vitro of human brain derived cells. An increased number of large, irregularly shaped residual bodies were found both in old cells which did not divide because they were in the terminal stage and in young cells kept in stationary phase. The latter cells retained the capacity for mitosis in spite of the increased amounts of residual bodies.

Collins and Brunk [1976] attempted to characterize the residual bodies formed in contact inhibited young cells. They were found to be acid phosphatase positive, to contain heavy metals, were acid fast with Ziehl-Neelsen and PAS-positive, had the characteristic fluorescence of lipofuscin pigment and were morphologically similar to the pigment. This reinforced the idea that the accumulation of residual bodies in cells aged in vitro is merely secondary to their inability to dilute their contents of residual bodies through division. Collins and Thaw [1983] also found that the autofluorescence per cell did not show fundamental differences between cells remain-

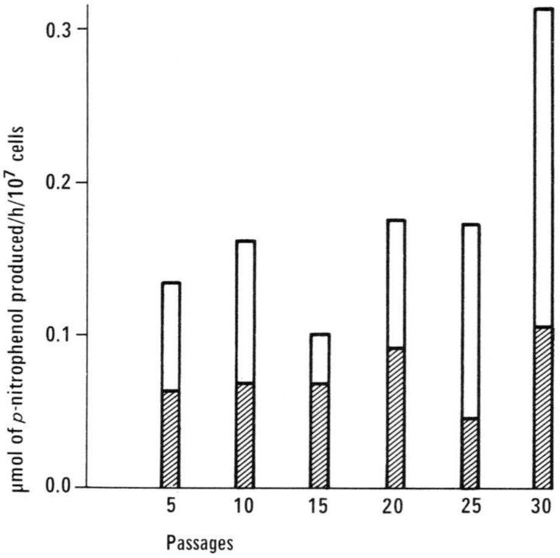

Fig. 32. Acid phosphatase activity at the time growth curves reached the plateau. The entire bars represent the total activity while the shaded areas indicate the free activity [Macieira-Coelho and Lima, 1973].

ing in G1 and those in S phase and G2. Although these investigators found supportive evidence for the lipid peroxidation origin of lipofuscin through the modulation of the amount of pigment in brain cells with various oxygen concentrations and exposure to pro- or antioxidants [Thaw et al., 1984], they questioned the idea that the presence of lipofuscin in itself denotes general peroxidation damage to vital cellular structures [Collins and Thaw, 1983]. It cannot be excluded, however, that the presence of these pigments has no secondary effects on nondividing cells.

Cytoplasmic autofluorescence was monitored by different groups to ascertain the presence of lipofuscin granules during in vitro aging. Deamer and Gonzales [1974] reported a progressive increase in the percentage of fluorescent cells during serial proliferation of human fibroblasts. Most fluorescent structures contained acid phosphatase and cells with high levels did not incorporate thymidine during a 24-hour period. The authors thought that these cells were nondividers although one knows now (see chapter IV) that they are slow dividers. This was indeed ascertained by Jonkind et al. [1982] who sorted out the autofluorescent cells and found that their population doubling time was increased 2.3-fold. The specific activities of lysosomal enzymes were also higher in these cells. The autofluorescent cells contained a large amount of small electron-dense granules that were classified as secondary lysosomes. The same group [Poot et al., 1985] found that the rates of increase of autofluorescent cells during in vitro aging was highest in cells from patients with Werner's syndrome. An interesting finding was the inhibition of the accumulation of autofluorescence by vitamin E in sorted nonautofluorescent cells, without affecting the formation of residual bodies. It was infered that the autofluorescence results from free radical reactions in lipid compartments of the cell, but that the formation of residual bodies is independent of this type of reaction.

Although vitamin E has this protective action, it has no effect on the aging in vitro of fibroblasts [Balin et al., 1977; Sakagami and Yamada, 1977]. This reinforces the idea that the accumulation of fluorescent pigments is not implicated in the mechanisms of senescence of dividing cells [Buchanan and Sidhu, 1986]. Another antioxidant, 2-mercaptoethylamine [Golczewski, 1984] does not seem to have any significant effect either, on the life span of cells in vitro. However, in adrenocortical cells vitamin E and selenium increased the replicative life span by 22–30 and 10–20%, respectively [Hornsby and Harris, 1987]. This seemed to be caused by the improved cloning efficiency caused by these two agents rather than by a protection of oxidative damage.

Pereira et al. [1976] approached the problem in an indirect way, studying the sensitivity to near ultraviolet and to visible lights of human embryonic lung fibroblasts (WI-38) at different population doubling levels. They found an inverse correlation between culture age and phototoxicity. Antioxidants (vitamin E and C and butyl hydroxytoluene) had a protective effect on the cells, from the light. Flavins seemed to be the main photosensitizers in visible light damage and the amount of riboflavin in the growth medium was directly related to the increasing sensitivity to light. These investigators could not find any difference in the amount of fluorescent peroxidation products between light treated cells and identical cultures kept in the dark.

The increased resistance of old cells to photosensitization, could be due to the increased superoxide dismutase (SOD) activity found in these cells [Duncan et al., 1979; Crosti et al., 1985]. Other investigators [Yamanaka and Deamer, 1974] did not find increased levels of SOD in human embryonic lung fibroblasts (WI-38) aged in vitro. This could be due to a tissue difference, since Duncan et al. and Crosti et al. used skin fibroblasts, or to a less sensitive assay to measure enzyme activity, utilized by Yamanaka and Deamer.

Duncan et al. [1979] also found a direct correlation between SOD levels and donor age in skin fibroblasts obtained from different human donors. So it seems that the old cells are at least as well protected as younger cells, from the production of free radicals. A feedback mechanism induced by the increased amount of material produced by lipid peroxidation may be responsible for the increased SOD activity. The material could accumulate through a disturbance of the mechanisms responsible for the removal rather than through an increased production.

The fact that different oxygen pressures do not change the life span of proliferating fibroblasts [Balin et al., 1977], also suggests that cells are well protected from the production of free radicals and that the limited life span is not due to increased peroxidation. Nevertheless, late passage cells, are more sensitive to hyperbaric oxygen [Honda and Matsuo, 1980], as estimated from growth curves and from the capacity to form colonies.

Some results, if confirmed, could suggest that the biochemical pathways involved in antioxidation, are modified when cells approach the terminal stage and could influence some of the characteristics of these stage. As mentioned in the chapter describing the final events, the ratio of reduced to oxidized glutathione decreases in human embryonic fibroblasts approaching phase IV [Takahashi and Zeydel, 1982]. On the other hand, the activity of gamma-glutamyl transpeptidase increases, an enzyme implicated in the degradation and resynthesis of glutathione. Furthermore, glutathionase and phosphate-independent glutaminase activities increase and so does ammonia production, a substance that has an influence on lysosomal function and is known to increase autofluorescence [Buchanan and Sidhu, 1986].

Poot et al. [1984], however, reported different results than those of Takahashi et al. [1982], working with skin fibroblasts from healthy subjects and from donors with Werner's and Spielmeyer-Vogt syndromes. These authors found a slight increase in reduced glutathione during in vitro aging and the cells from the patients did not differ in this respect from those of the normal ones. Poot et al. used a different methodology and different cells; both or either of these could be responsible for the opposite results. Further experiments would be necessary to clarify this point. The reduced glutathione recovery rates after depletion with diamide, was found to be identical in fibroblasts from healthy subjects and from patients with Werner's and Spielmeyer-Vogt syndromes [Poot et al., 1986].

In WI-38 fibroblasts the concentration of reduced glutathione increases suddenly in the postmitotic state; glutathione peroxidase activity on the other hand, decreased already at early population doubling levels [Funfu et al., 1985]. Glutathione reductase activity decreased at a constant rate and reached a drastic low level in the postmitotic state [Funfu et al., 1985].

Hornsby et al. [1984] tested the free radical hypothesis on differentiated cells with a limited life span, i.e. bovine adrenocortical cultures. Although they concluded that the finite culture life span of normal diploid adrenocortical cells does not result from oxidative damage, their results show that antioxidants can intervene in regulatory pathways and protect the cells in some regards. Hornsby et al. noted that during serial divisions of adrenocortical cells there is a loss of inducibility of 11-beta-hydroxylase. However, when the experiment was performed in a lowered oxygen tension or in the presence of antioxydants, there was no loss of reinducibility. So it seems that antioxidants can stabilize the differentiated properties of these cells [Horsnby, 1980]. Furthermore, adrenocortical cells show during the course of their in vitro life span an increasing sensitivity to aminooxyacetate [Hornsby, 1982]; vitamin E and selenite could completely prevent cell death caused by aminooxyacetate [Hornsby, 1982]. Recently, Hornsby and Harris [1987] reported that DNA cross-links and strand breaks, two types of damage induced by an oxidative damaging agent, were not prominent either in early passage or presenescent cultures.

Thus, although the accumulation of oxidation products does not seem responsible for the limited life span of cells, it seems that these metabolic pathways intervene in the regulation of other cellular functions. In other words, the phenotype of cells, other than that related to the potential number of divisions, may vary depending on oxidative reactions.

Indeed, there is a tendency to look upon the consequences of free radical reactions in a negative way, i.e. in terms of degenerative processes. However, these biochemical pathways must be very important in physiological processes, and as mentioned above, in the protection of cells from external events. In this regard, Croute et al. [1986] explained the prolongation effect of low dose rate radiation, on the division potential of human fibroblasts, by the intracellular enzymatic scavenging of radioproduced H_2O_2 generated either by OH recombination or by chemical or enzymatic dismutation of O_2^-.

Thus, one can conclude by saying that the problem is not settled and that the influence of oxidation products on other cellular functions evolving during serial divisions and on the response of cells to environmental factors should be searched for. Protection from oxidative products, although not of primary importance for the growth potential, may allow the preservation of more subtle cellular functions. It should not be forgotten that the production of free radicals is part of the electron transport chain where electric energy is transformed into chemical energy. Thus, the real role of free radicals will probably be known only when the regulation of these processes of energy transduction in the cell will be understood.

e. Mitochondria

The hypothesis implicating mitochondria in the mechanism of cellular senescence is related with the free radical theory. Indeed, it has been proposed [Miquel and Fleming, 1984] that the fundamental cause of cell aging is a lack of balance between mitochondrial DNA repair and the disorganizing effects of oxygen radicals. This view was expressed within the concept of aging as the result of cell differentiation. Since differentiation would require higher levels of O_2, it would result in the formation of free radicals whose products would then damage mitochondrial DNA.

Mitochondrial DNA (mtDNA) copy numbers, methylation and its integrity, have been studied in normal and mutant skin fibroblast cultures [Shmookler Reis and Goldstein, 1983]. The number of mtDNA molecules per nuclear equivalent was determined hybridizing total cellular DNA with pure mtDNA and with cRNA copied from mtDNA. Methylation and integrity of DNA was evaluated from the relative proportions and mobilities of restriction fragments. The data indicated that neither the number of copies, the integrity nor methylation pattern of mtDNA is substantially altered during serial passages of these cells. A significant heterogeneity was observed though, between individual fibroblast clones. White and Bunn [1985] could not obtain either, any evidence suggesting the presence of deletions, insertions or rearrangements in human mitochondrial DNA during in vitro aging.

The specific activity of DNA polymerase-gamma, which is now considered to be the mitochondrial DNA polymerase, also remains constant during in vitro aging of human embryonic lung fibroblasts [Murray, 1981]. Thus, this aspect of mitochondrial replication is not affected by in vitro senescence.

Some modifications, however, must take place in mitochondria, as judged from morphological studies. Several investigators observed profound changes in the morphology of mitochondria under electron microscopy, which concerned essentially bizarre shapes and a switch in the orientation of the christae which became longitudinally oriented instead of the transverse orientation seen in the young cultures [Brock and Hay, 1971; Lipetz and Cristofalo, 1972; Johnson, 1979a; Basler et al., 1979; Goldstein et al., 1984]. The mitochondrial matrix appeared more condensed and the cytoplasm contained less mitochondria. Dilated christae were also observed and cystic blebs were evident in mitochondria of some cells aged in vitro and from Progeria patients. The latter and the decreased mitochondrial width seen with high voltage electron microscopy could mean an age-dependent loss of inner membrane integrity [Goldstein et al., 1984].

The uptake of the dye rhodamine-123 which is selectively taken up by mitochondria of living cells [Johnson et al., 1980] was used as a probe for the presence of modifications at the level of the mitochondrial membrane. When no measurements were made [Goldstein and Korczack, 1981] no gross derangements in the morphology or number of mitochondria could be detected. When measurements were made, however, human skin fibroblasts aged in vitro were found to bind more dye and exhibit two distinct peaks of fluorescence intensity detected with a fluorescence activated cell sorter suggesting the presence of two subpopulations [Martinez et al., 1986].

Another result suggestive of an alteration in these organelles, is the 300% increase in the specific activity of cytochrome oxidase through the life span of WI-38 fibroblasts, which is a marker enzyme for the inner mitochondrial membrane [Sun et al., 1975]. Hakami and Pious [1968], however, did not find an increase in this enzyme in in vitro aged human fibroblasts derived from newborn foreskin. The opposite results could be due to tissue differences. Remacle et al. [1980] measured cytochrome oxidase activity on a mitochondrial fraction obtained from WI-38 cells by differential centrifugation and isopycnic equilibration in a density gradient. Compared to the cell homogenate, there was an 8-fold increase in specific activity for cytochrome oxidase in the peak fraction.

The experiments of Hornsby described in the preceding section, where the relevance of oxidative damage for the survival of adrenocortical cells was studied, have also implications for the problem of the mitochondrial role in cellular senescence. The loss of the inducibility at 19% O_2 tension of the enzyme 11-beta-hydroxylase, a mitochondrial cytochrome P-450, indicates changes taking place at early culture stages in mitochondria [Hornsby, 1980]. The rise in the ratio of oxidation of glutamine and pyruvate, points in the same direction and so does the increased susceptibility to the toxic effects of aminooxyacetate, an inhibitor of aminotransferases [Hornsby et al., 1984]. The latter seems to be toxic because it prevents utilization of glutamine in mitochondrial energy production [Hornsby, 1982]. These mitochondrial related changes, may have implications for some of the functions of this cell system but do not seem relevant for their proliferative potential [Hornsby et al., 1984]. They are of importance for senescence of these cells since the metabolic modifications accompanying proliferation are part of the aging process, although not implicated in the cell doubling potential.

The respiratory and glycolytic capacities during serial proliferation in vitro of human embryonic cells, remain constant [Cristofalo and Kritchevsky, 1966]. The comparison of fibroblasts from human donors of different ages, however, revealed a decline in glucose and glutamine oxidation, which require mitochondrial function [Peterson and Goldman, 1986]. The same authors reported that these processes are also more altered by Alzheimer's disease than by normal aging. Goldstein et al. [1982] on the other hand concluded that there are no gross deficits in energy metabolism during aging of human postnatal fibroblasts in vitro. They found, however, an increased dependence on glycolysis of old cells during the period of growth. This was interpreted in terms of increased energy needs and/or less efficient oxidative phosphorylation. Goldstein et al. [1982] also studied progeria cells which were equal to normal cells in O_2 consumption. They differed though from the controls by an increased lactate production during the growth period, a feature also shared by normal cells aged in vitro.

The electrophoretic mobilities of mitochondrial proteins from fibroblasts from ageing syndromes were identical to those obtained from fibroblasts of healthy subjects [Yatscoff et al., 1978].

These, at the moment, limited data concerning the role of mitochondria in senescence of dividing cell compartments, seem to reveal a path worth pursuing. As we mentioned in the introduction, the proliferative potential is just one parameter among others to study how a cell evolves through division. Other metabolic processes are equally important since they must be part of the composite picture that constitutes the aging syndrome. The data suggest that some metabolic processes conditioning the cell response to external stimuli are affected by modifications occurring in mitochondria and hence that these organels should not be neglected. In particular, they may be implicated in the maintenance of some differentiated properties as suggested by the experiments of Hornsby.

IV. Dynamics of the Loss of the Proliferative Potential during Cell Senescence

The need to understand what happens during the division cycle when cells lose their growth potential led to interesting findings which shed some light into the complexity of the events that influence cell proliferation. This did not take place without controversy and misunderstandings and could constitute an example of how concepts evolve in the scientific community and of the effort needed for scientists to abandon ideas deeply rooted in the cultural background. Actually, what took place in the field was an opposition between those that introduced the concept of uncertainty to the kinetics of cell proliferation, and the ones that used a dualistic interpretation, i.e. considered the cell population was constituted just by dividers and nondividers. The latter attitude is a cultural aspect of the western civilization, which tends to have a Manichean approach for the interpretation of different phenomena.

Theoretically, the proliferation of cells during exhaustion of their division potential, could evolve in three ways [Macieira-Coelho et al., 1966a]: (a) a progressive shift of the population to a slow uniform rate of proliferation; (b) a decline of the fraction of the dividing cell population with the rest of the cells unable to divide, and (c) an increased heterogeneity of the population showing a spectrum between the two extremes, i.e. complete inhibition and a normal division cycle.

The first attempts to understand the dynamics of the proliferative changes during the loss of the division potential [Macieira-Coelho et al., 1966a, b; Macieira-Coelho, 1967a, 1973a, 1974; Macieira-Coelho and Berumen, 1973] concluded in favor of the third model, i.e. a spectrum of cells with different probabilities of cycling. It was found that at each doubling, some cells enter division once, others two or three times, and still others do not divide between subcultivation and resting phase but will do it during the following passage. The size of these different cell compartments varies with the number of cells inoculated. Moreover, the cycling cells have a spectrum of generation times from short to long cycles; the exact fraction of nondividing cells was difficult to evaluate. These works

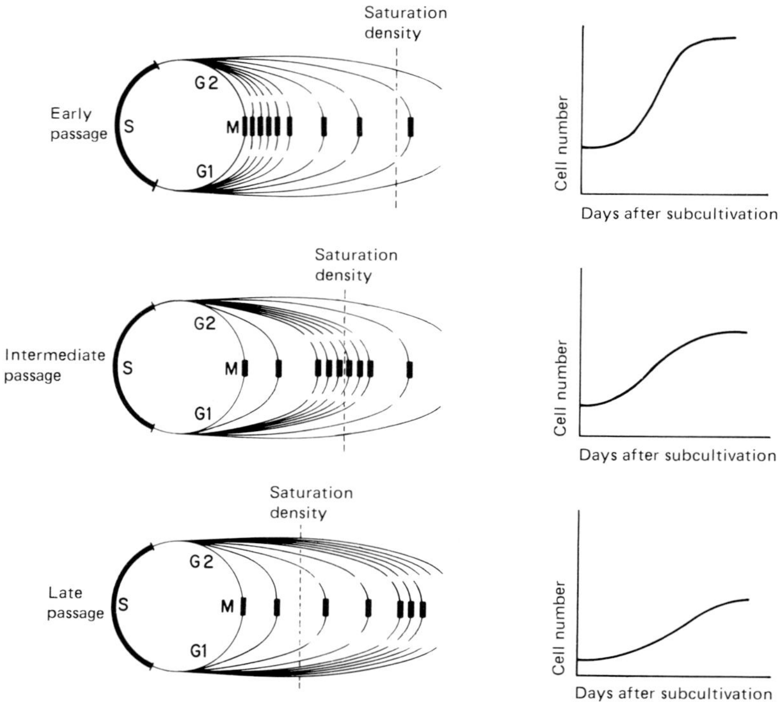

Fig. 33. Schematic representation of the kinetics of proliferation of human fibroblasts in vitro at 3 different population doubling levels. Each circle or ellipse represents the generation time of a cell. The vertical dashed lines correspond to the saturation density. On the right hand side are represented the growth curves that are obtained at the corresponding doubling levels.

introduced the notion of uncertainty to the probability of cycling and to the nondividing state.

Based on these results, a model of the dynamics of cell proliferation during senescence of the fibroblast stem cell compartment, as illustrated in figure 33, was proposed [Macieira-Coelho, 1977]. Three different periods of the cell population life span are represented. On the left side of the figure each circle and ellipse corresponds to the generation time of a cell. At an early stage most cells in the population are postulated to have short division cycles. Some cells already exist with long generation times (ellipses) which are extended mainly at the expense of the G1 and G2 periods. There is no gross prolongation of the S period and mitosis and the saturation density is high (vertical hatched line). At this stage the growth curve of the population (right hand side of the figure) has a short lag period, a phase of logarithmic growth with a steep slope, and a high saturation density reached within 3–4 days. Towards the middle of the population life span, there are still cells with short generation times, but many cells have division cycles between the two extremes. At this stage the slope of the logarithmic portion of the growth curve will be less steep and the cells reach a lower saturation density, although still within 3–4 days. Since the cells become more sensitive to cell cycle inhibition during crowding (vertical hatched line) more cells will slowdown during the G2 period.

Towards the end of the life span, although most cells have very long generation times, there are still cells with short cycles. The growth curve will be characterized by a long lag phase, a period of short logarithmic growth with a shallow slope and a low saturation density reached only after 7–10 days. In the figure the prolongation of G1 and G2 is identical but it is possible that one is predominant. Most of the cells that were first thought to be blocked in G2 [Macieira-Coelho et al., 1966a] were later found to be delayed in this period of the cell cycle [Macieira-Coelho, 1967a; Macieira-Coelho, 1973b; Macieira-Coelho and Berumen, 1973]. Some cells, though, seem to be blocked in G2 [Kapp and Klevecz, 1976].

It should be emphasized that when one speaks about prolongation of the G1 and G2 periods, it applies only to cells that are moving through the cycle. Since the cells that are arrested due to the effect of saturation density, are mainly in G1, a distinction should be made when measuring the prolongation of the G1 and G2 periods, otherwise cells that are arrested will be taken as prolonged in G1. Methods that allow the analysis of cells progressing through the cycle should be distinguished from those measuring only instants [Macieira-Coelho, 1973b]. This concept of measuring the transit of cells is more important for aging than the evaluation of irreversible arrested cells which has misdirected several investigators. Indeed, the in vivo aging of dividing cell compartments is not a question of accumulation of irreversibly arrested cells, but rather a deregulation of the response to proliferation (see chapter II.3).

Subsequent data, however [Merz and Ross, 1969; Cristofalo and Sharf, 1973; Smith and Braunschweiger, 1979; Ban et al, 1980a; Pontén et al., 1983; Karatza and Shall, 1984], were interpreted in terms of the second model, i.e. that at each doubling the nondividing cell compartment increases exponentially reaching finally a stage where the whole population is nondividing. This dualistic interpretation is illustrated in figure 34

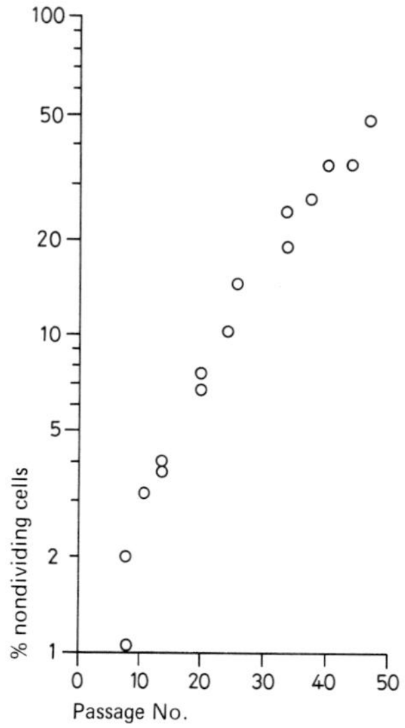

Fig. 34. Summary of repeated determinations of the fraction of nondividing cells evaluated from the cloning of a single WI-38 culture from passage 8 to passage 47 [Merz and Ross, 1969].

which shows the number of cloning forming cells of a human fibroblast population at different PDL. It was concluded that the cells that did not give rise to clones had lost the ability to divide and that the cloning experiments expressed the events in the culture at the population level. Merz and Ross [1969] considered that aging at a cellular level could be defined by the increase, with advancing age, in the probability that an individual will not survive the next unit of time. This view was reinforced by the results of Cristofalo and Sharf [1973] measuring dividing cells with ^{3}H-thymidine, and has strongly dominated the field of cytogerontology. Later, Merz and Ross [1973] conceded that the clones obtained are heterogeneous in the number of cells generated and that the heterogeneity increased with time after plating.

It was later shown that this type of approach was analysing the events only in terms of what happens to the most vigorous cells in the population [Macieira-Coelho and Taboury, 1982; Macieira-Coelho and Azzarone, 1982]. The investigators that adhered to the dualistic interpretation, ignored that the commitment to divide depends on several factors, inter-alia, the nature of the substratum for cell attachment [Macieira-Coelho and Avrameas, 1972; McKeehan and Ham, 1976], the concentration of nutrients and growth factors in the medium and the co-operation between cells, either through secretion of regulatory substances and creation of a microenvironment favourable for division [Eagle, 1959], or through contact stimulation of growth [Ohno and Yamada, 1984]. This latter aspect is disrupted during cloning conditions where by definition one attempts to measure the number of cells capable of dividing when isolated. In fact, it is well known that one can improve cloning efficiency either with enriched media [McKeehan and Ham, 1976] or with a feeder-layer of irradiated cells; Cox and Masson [1974] increased the number of clonogenic human fibroblasts from 5 to 60% by adding a feeder-layer of irradiated cells. Hence it appears that cloning efficiency measured with the experimental conditions used, will reveal only the most vigorous fraction of the population.

Moreover, the works that reached the same conclusion as Merz and Ross measuring the number of dividing cells with ^{3}H-thymidine used a concentration of the precursor which is growth inhibitory [Pollack et al., 1979]. Some investigators believe that the precursor will eliminate only the cells capable of synthesizing DNA so that the remaining nonlabeled cells are nondividers. This is again a misunderstanding of the various factors that can influence the commitment to divide. As mentioned above, the latter does not depend only on the intrinsic properties of the cell; one of the parameters that influences the triggering of the division cycle is cell cooperation. Eliminating the most vigorous cells with the radioactive precursor stops them from secreting growth stimulating factors and nutrients that could induce neighboring cells to divide. It will also stop them from migrating, thus decreasing the chances of contact stimulation of growth. Experiments performed on palladium islands where one can maintain cells isolated from each other but still have the cell population in the same nutrient medium, showed that the number of dividers increases if the cultures are started with two cells per island instead of one [Blomqvist et al., 1987]. The number of dividers when doublets are present, increases even more if the nutrient medium is supplemented with a higher serum concentration or with $FeCl_2$. These experiments show that some cells can initiate the division cycle with a small stimulus and that other cells need a stronger stimulation. Hydrocortisone which can stimulate and prolong the life span of human embryonic lung fibroblasts [Macieira-Coelho and Loria, 1974] can also induce DNA synthesis in cells which otherwise would seem to be nondividers [Phillips et al., 1982]. Furthermore, the proportion of noncycling cells changes as a function of mitogen concentration [Rabinovitch, 1983]. The loss of the response of cells during in vitro senescence to various growth factors was found to be progressive and selective for the different factors [Tsuji et al., 1984]. At the very end, 100% life span completed, only SV_{40} virus could induce DNA synthesis [Tsuji et al., 1984]. All these works show that during cell senescence the first manifestation is not the increase of nondividing cells but rather the need of an increased stimulus to initiate the division cycle.

Absher et al. [1974] and Absher and Absher [1976] used an elegant approach to analyse the descent of single cells without suppressing the interactions with the whole population. They isolated groups within small ponds in a Petri dish, so that the supernatant was still the same for the whole population in the dish, and followed with time-lapse cinematography the genealogy of individual cells. They could thus compare clones obtained from different PDL concerning interdivision times, number of generations and clonal size up to confluency. They found at higher PDL, prolonged generation times and an increased heterogeneity of interdivision times. The division potential of individual cells within the same PDL was different since each cell could produce a different number of descendants due to variation of interdivision times and the appearance of cells that did not divide. The number of nondividing cells was around 10% until the 40th doubling and increased only afterwards, a finding which agrees with the data obtained by Vincent and Huang [1976]. Hence, Absher's observations favored the concept of heterogeneity and uncertainty of the probability of cycling and of the non-dividing state (fig. 35).

One of the interesting results of the work of Absher et al. [1974] was the finding that the correlation coefficient of interdivision times of sister pairs decreased with age and that cells with a preceding long generation time and large size could give rise to small daughter cells with shorter generation times. Smith and Whitney [1979] also found a difference in the ability to proliferate between the two cells arising from a single mitosis. They could differ by as much as eight population doublings, i.e. 256-fold in the number of cells produced. They also observed the rapid development during serial divisions, of intraclonal variability in doubling potential. Absher's experiments also revealed fluctuations between cycling and resting stages suggesting a periodicity in the behavior of cells. Reversion was also suggested in the work of Mitsui and Schneider [1976a] who separated the large, slowly dividing cells from the small, rapidly dividing ones

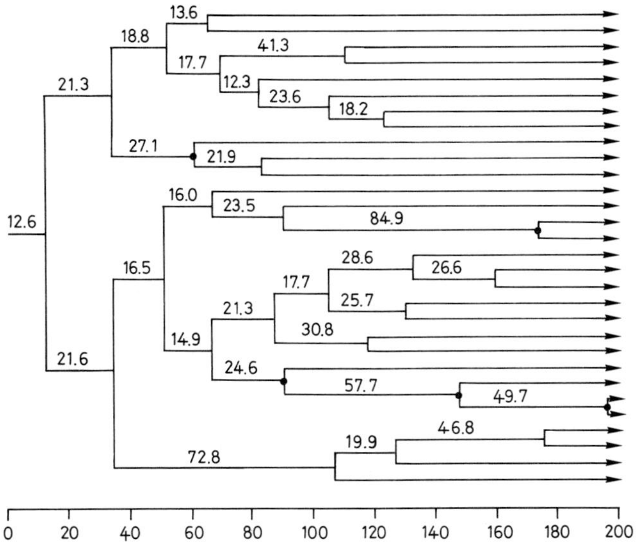

Fig. 35. Pedigree of a clone obtained from human embryonic lung fibroblasts at the 51st population doubling level. Numbers on lines indicate interdivision time in hours. ● = Large cells [Absher and Absher, 1976].

and found a shift from one compartment to the other. However, since a 100% separation of each type of cell is impossible, reconstitution of a compartment by residual cells cannot be excluded.

These findings of reversion into the division cycle are reminiscent of the experiments of Dell'Orco et al. [1974] who were able to prolong the doubling potential of human fibroblasts by keeping the cells for a long period of time in a near nonmitotic state. When growth resumed in the arrested population, the fraction of dividing cells was also increased [Dell'Orco et al., 1973]. One might wonder whether the prolongation of the life span and the higher fraction of dividing cells after long periods in resting phase is due to some sort of regeneration taking place during a reduction of the demand for cell renewal. The same reasoning could explain a phenomenon that occurs after serial transplantation in vivo of haemopietic cells [Vos and Dolmans, 1972] and of mammary glands [Daniel and Young, 1971] where the number of possible transfers increases with longer intervals between transplants. Pertinent in this regard are the experiments of Inoue and Cronkite [1983] who could remove some of the proliferative defect of spleen colony-forming units from old mice by in vivo incubation.

Several authors used Absher's data to test a stochastic model concerning the transition into the cell cycle during exponential growth. This model, first described by Burns and Tannock [1970] and later reappraised by Smith and Martin [1973], considers one single exponentially distributed random distribution. Some authors [Grove and Cristofalo, 1976] claimed that old, but not young cells deviate from the transition probability model, others on the contrary considered that cell behavior can be described by the model at any PDL, the observed discrepancies being due to changes in probability and minimum interdivision times [Smith, 1977]. The discrepancies depended on whether they used all the cells produced during the experiment [Grove and Cristofalo, 1976] or tried to render the population more homogeneous by selecting for cells with identical growth rates [Shields, 1979] or generations [Smith, 1977]. According to Mets and Verdonk [1978] the deviation from the model is due to increasingly long generation times during the observation period after the first three generations. More recently, it has been suggested that differences in cell metabolism would be a more reasonable explanation for the differences in cycle times rather than random transition [Castor, 1980].

The fact that a plot of the percentage of cells that did not divide against time was rather similar for the first generations at different PDL confirms the findings of Kapp and Klevecz [1976] and Macieira-Coelho and Azzarone [1982] which show that at any age of the population there is a fraction of actively dividing cells with the same kinetics of proliferation. This is obvious in experiments using bromodeoxyuridine label followed by Hoechst and Giemsa. By the pattern of chromatid staining it can be determined, which cells in mitosis have done 1, 2 or 3 divisions between the time of labeling and the time of fixation. Figure 36 illustrates such an experiment showing that through the whole life span, cells making 2 and 3 divisions within 48 h, can be found. The descendants of these cells, however, seem to be differently affected at higher PDL by all the environmental parameters that influence the capacity to enter the division cycle. Events such as cell proximity, degree of asynchrony and depletion of nutrients will also influence the capacity of aged cells to enter the cycle.

As a matter of fact the different interpretations show the impossibility of defining by a simple probability law such a complex and unpredictable behavior of this cell system in regard to the kinetics of proliferation [Skehan and Friedman, 1979].

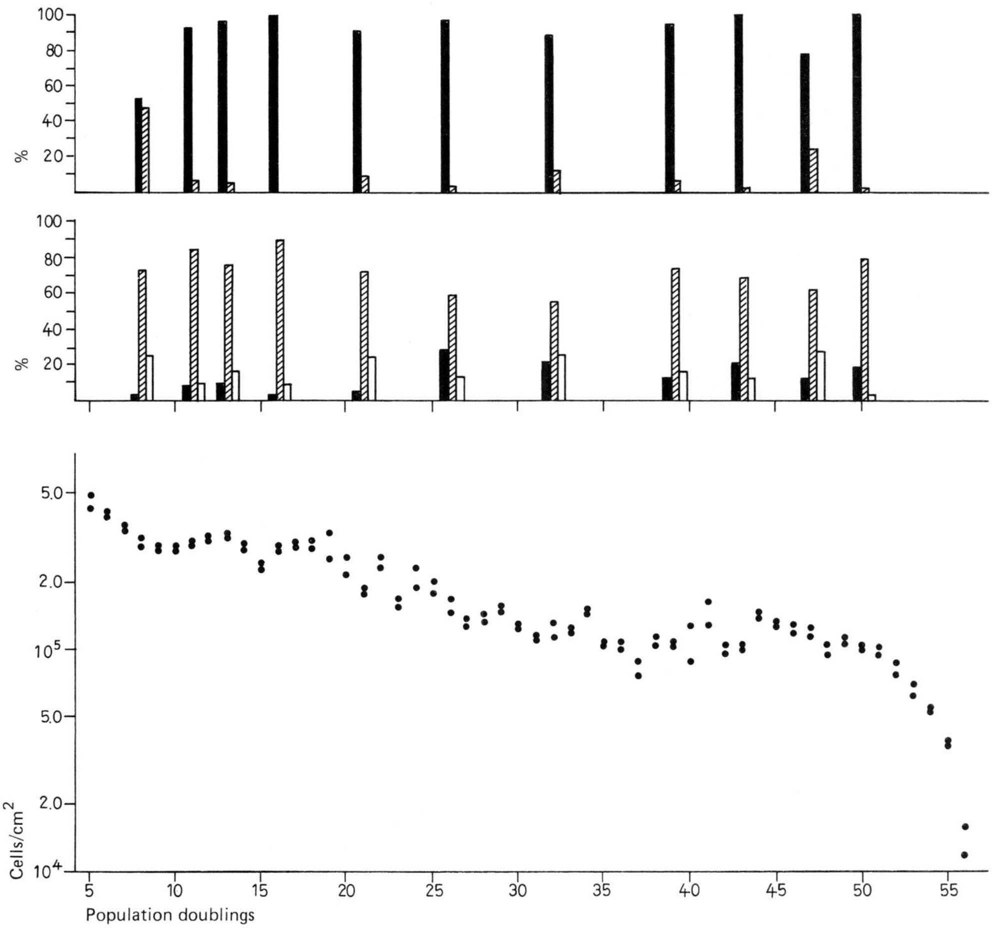

Fig. 36. Maximal cell densities (●) found at each population doubling of human embryonic lung fibroblasts. Each circle corresponds to one cell count. The columns indicate the % of mitoses corresponding to cells that made one (■), two (▨) and three (□) divisions after 24 h (upper row) and 48 h (lower row) labeling with bromodeoxyuridine at different population doubling levels [Macieira-Coelho and Azzarone, 1982].

One of the important implications of the studies of the kinetics of proliferation of the fibroblast cell system is to have led to the notion of the uncertainty of the nondividing state [Macieira-Coelho, 1974, Macieira-Coelho and Taboury, 1982].

In conclusion, senescence of these dividing cell compartments consists of a progressive shift to a new equilibrium, where the main events are not the abrupt creation of an irreversible new state, but rather a change in the probability for a cell to be in a given state.

Several experiments showed that the changes in the kinetics of proliferation follow a progressive pattern and reach a final stage with abrupt events [Macieira-Coelho and Azzarone, 1982]. This is apparent from the maximal cell densities reached at each PDL (fig. 1), from the percentage of cells initiating DNA synthesis after attachment and spreading, and from the maximal percentage capable of synthesizing DNA during a 24-hour period before resting phase (fig. 26). Other methods that can detect nondividing or very slowly dividing cells support the finding that cells with a very low probability of cycling or postmitotic ones accumulate only at the very end. Indeed, the fraction of autofluorescent fibroblasts [Jongkind et al., 1982] and of red blood cell binding to lectin-coated fibroblasts [Aizawa et al., 1980a], which are characteristic of non-dividing cells, increase only during the last few doublings. Although some terminal cells are produced at each doubling [Rao, 1981] most must be lost.

Even at the very end it is difficult to define in a mass population, which cells become irreversible nondividers since a continuous labeling with ^{3}H-TdR shows that the number of labeled cells increases progressively to close to 100% [Macieira-Coelho, 1974; Schneider and Fowlkes, 1976; Ryan, 1979b]; as the radioactivity eventually eliminates the labeled cells, the decline of nonlabeled cells must be due to the recruitment of new cells and not to dilution. Furthermore, labeling with BrdU, which eliminates dividing cells, does not eradicate completely the latter even after several days [Burmer and Norwood, 1980] and changing the nutrient medium repeatedly to terminal cells, induces a population doubling although it can take as long as 4 months [Pereira-Smith and Smith, 1981].

This type of evolution is found in embryonic as well as postnatal human fibroblasts, although the latter seem to have already, early during their life span in vitro a higher heterogeneity and a higher fraction of slow dividing cells [Macieira-Coelho and Pontén, 1969; Vincent and Huang, 1976].

Fibroblasts from other species also seem to behave, basically in the same way, concerning the kinetics of proliferation, although the heterogeneity can assume different degrees and the timing of the events may be contracted or expanded depending upon the potential number of doublings and upon other cell properties.

In chicken fibroblasts for instance, close to 100% of cells synthesize

DNA during most of the life span (approximately 35 doublings) and at the very end, sudden events slow down dramatically the capacity to divide [Lima and Macieira-Coelho, 1972]. In fibroblasts from this species it was also found that an increase in interdivision times and not an accumulation of nondividing cells explains the proliferation kinetics of old cells since the latter can achieve the same percentage of labeled nuclei when given a longer pulse period [Ryan, 1979]. Furthermore, up to 78% of an old cell population with 94% life span completed, can resume active proliferation when inoculated and propagated under more favorable growth conditions [Ryan et al., 1982]. The fact that in chick fibroblasts cultures only 4% of the clones are capable of more that 35 doublings [Beug and Graf, 1977] gives an idea of the extreme heterogeneity of these cells and how a completely different picture can be obtained depending on the methods used to analyse proliferation kinetics. Chick chondroblasts, retinal pigment cells [La Rocca and Rafferty, 1982] and articular chondrocytes [Dominice et al., 1986] also have the same type of kinetics of proliferation as fibroblasts.

In mouse cells the decline in growth potential also takes place by a spread in the spectrum of probabilities of dividing [Preumont et al., 1983]. In this species the chronology of events is contracted because of the short division potential. With mouse cells, however, another phenomenon is superimposed, as is discussed in chapter II.2, because of the propensity of these cells to overcome the growth arrest, and immortalize.

Although the kinetics of proliferation of other cell types has not been studied in such detail, at least another one, aortic endothelial cells, seems to behave very much the same [Rosen et al., 1981].

V. Analogy between Changes in the Kinetics of Cell Proliferation and Metabolic Events

Hayflick and Moorhead [1961] described three phases during senescence of human fibroblasts in vitro. From the works described above, Macieira-Coelho and Taboury [1982] proposed the existence of a fourth, short, terminal phase. The existence of phase I is suggested from the initial rise of the saturation densities (fig. 1), from the morphological observations showing the appearance of a more homogeneous population, and from chromosome studies [Miller et al., 1977] that detected initial changes in cultures of embryonic lung fibroblasts which disappear during the first passages.

Although not all parameters analysed remain steady during the following subcultivations, phase II can be defined as the period where the fraction of nondividing cells recorded by autoradiography is relatively constant at each doubling and not larger than 5–10%. For the sake of simplicity one can distinguish three states in phase II populations: a quiescent state in which cells permanently remain or periodically enter, corresponding to a small fraction of the population; a second state characterized by high proliferative activity and containing a progressively decreasing number of cells (fig. 1, 26); and a third state corresponding to cells with more heterogeneous behaviour. Phase II would evolve through an equilibrium between the two latter states, with a progressive predominance of the more heterogeneous one. The heterogeneity is manifested at the start of the cell cycle, in the time needed to complete the cycle, and in the sensitivity to the mechanisms which arrest cells due to proximity.

Phase III would be characterized by the slight increase in the number of cells which do not divide during a 24-hour period and by the prolongation of the population doubling time.

The existence of a final phase IV, corresponding to the last three or four doublings, is apparent from the rapid fall of the maximal cell density (fig. 1) and of the percentage of cells synthesizing DNA during a 24-hour labeling with ^{3}H-thymidine (fig. 26). Several metabolic events occur during the last 3–4 doublings, which support the idea of a final phase with distinct characteristics.

At the molecular and supramolecular level, modifications take place which follow the same pattern as the kinetics of proliferation described above, in other words, some events evolve progressively through phases I, II, and III and others occur abruptly during the last 2–3 doublings.

1. Progressive Events

It is obvious from the evolution of the kinetics of proliferation that something must take place during cell division, which progressively changes the cells, and, finally, create the conditions for the final abrupt steps. One change that evolves progressively during phases I, II and III and finally assumes an abrupt pattern during phase IV, is the genome reorganization described in chapter III.1. This reorganization could be a random or a programed event leading to the development of well-defined cell types. In any case, as described above, it can explain several of the features occurring in the division cycle and many of the metabolic modifications originated during serial divisions.

Another progressive change occurring in human fibroblasts is the increase in cell, nuclear and nucleolar dry masses and areas [Mitsui and Schneider, 1976a, b; Bemiller and Miller, 1979]. The patterns of increase were similar for all measures; the increases in nuclear and nucleolar areas were interdependent until the last doubling where they increased independently. The chronology of the increase in cell volume was monitored by Lima and Macieira-Coelho [1972] in chicken fibroblasts and compared with other events. The change in size followed a three-step curve (fig. 37); the first increase (10th population doubling) coincided with the change in the time needed to reach the maximal DNA synthesis after subcultivation (fig. 38), then there was a final increase before phase IV. It suggests that the modifications in cell volume per se are crucial in the increased difficulty in the initiation of the cell division cycle. The first event that could be detected in this cell system was a decline in protein synthesis on a per cell basis, which preceded the first increase in cell size (fig. 39). Protein synthesis per cell declined progressively through the population life span. Then at the 15th doubling there was a decline in total RNA synthesis per cell which coincided with a prolongation of the population doubling time. This was followed after the 20th doubling by a decline in the satura-

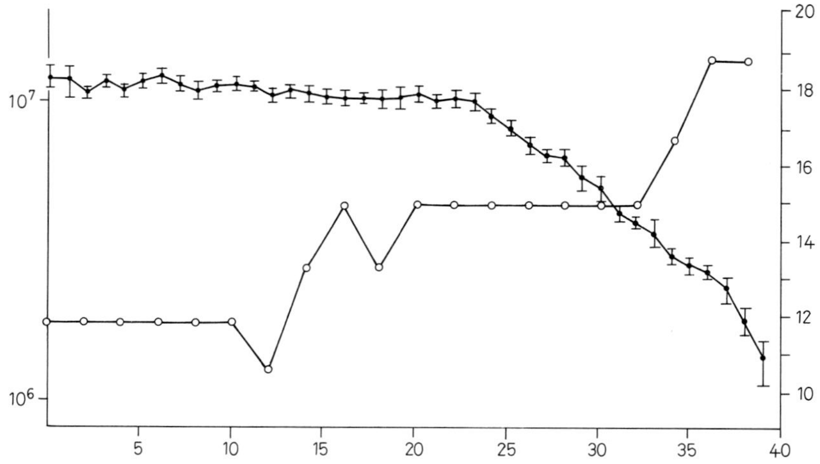

Fig. 37. Abscissa = number of passages; ordinate, left = number of cells/culture; right = cell size in μm. Modes of the distributions of cell volumes (○) plotted with the saturation densities (●) at each passage of chicken embryonic fibroblasts. The points on the curve are the mean of four cultures counted before each subcultivation and the vertical bars indicate the extreme values found within samples [Lima and Macieira-Coelho, 1982].

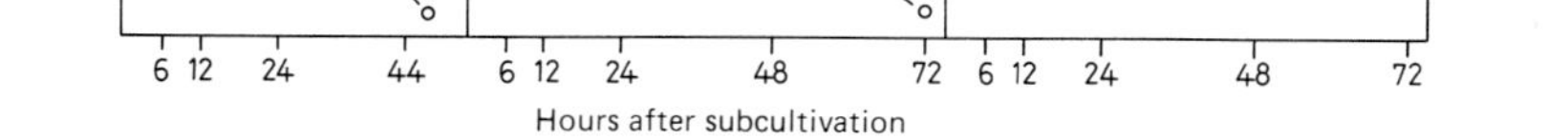

Fig. 38. Growth curves (—) and ^{3}H-TdR incorporation (----) after pulse labelings at different times after subcultivation. The numbers within parentheses indicate the passage when the experiment was performed. The respective doubling times (DT) are indicated. At the 15th passage the maximal DNA synthesis activity is found 12 h after subcultivation instead of 6 h as before, coinciding with the first increase in cell volume; after the 35th passage the second increase in cell volume coincides with an increase in the doubling time [Lima and Macieira-Coelho, 1972].

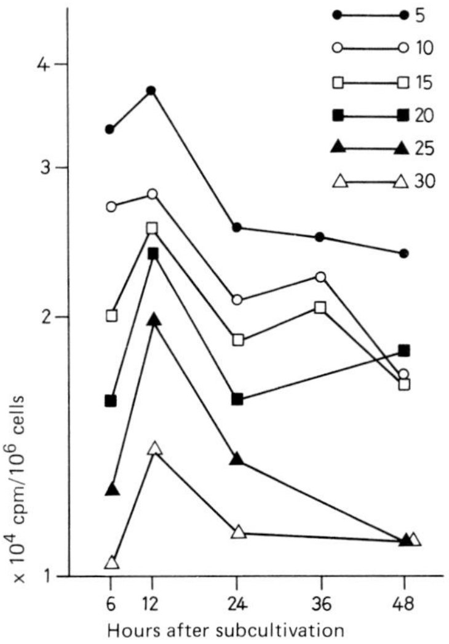

Fig. 39. Amino acid incorporation after pulse labeling chicken embryonic fibroblast cultures at different times between subcultivation and the time the growth curves reached a plateau. The passages at which the experiments were done are indicated [Macieira-Coelho and Lima, 1973].

tion density and a further decline in the total RNA synthesis. Finally, in phase IV, there was a higher activity of acid phosphatase (fig. 32). The relationship between RNA content, cell volume and growth potential was also studied by Whatley and Hill [1979] in human embryonic fibroblasts. Cell volume changes seemed to be related with the decline in the rate of cell division and an increase in RNA content.

An interesting system that has been studied in regard to the progressive changes occurring during aging in vitro are bovine adrenocortical cells [Hornsby et al., 1984] (fig. 40). The concentration of ACTH, necessary to stimulate steroidogenesis, becomes higher towards the middle and the end of the cells life span and there is a loss of ACTH-stimulated cyclic AMP production. Moreover, there is a change in the conditions for reinduction of 11-beta-hydroxylase by agents that raise intracellular cyclic AMP. The response to prostaglandin E1, angiotensin II and fibroblast growth factor is maintained. The ratio of oxidation of pyruvate/glutamine rises during the early part of the cells' life span and then remains steady. Up to the end of the cell population life span, inducible steroido-

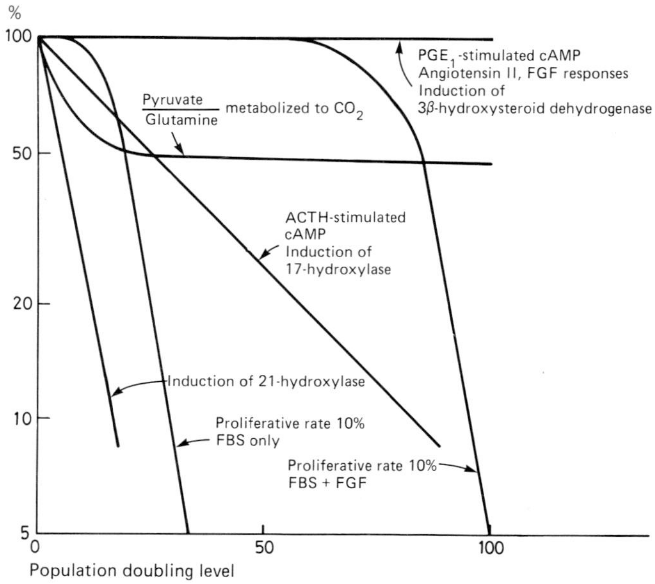

Fig. 40. General scheme of the life history of bovine adrenocortical cells in culture. The ordinate shows the measured level of the indicated functions after various numbers of divisions in culture, taking the level in the primary culture as 100% [Hornsby, 1984].

genesis is maintained but there is a loss of the synthesis of the normal final product, i.e. cortisol [Hornsby et al., 1979].

In the fibroblast system a progressive event that is directly related with the engagement into the division cycle is the decline in the ratio between the activities of two enzymes of the purine salvage pathway, hypoxanthine-guanine transferase and adenine-phosphoribosyltransferase [Paz et al., 1981]. In regard also to the problem of the availability of the DNA building blocks, there is a progressive decline in ribonucleotide reductase and in the resistance to hydroxyurea [Dick and Wright, 1982] (fig. 41, 42), and variations in deoxyribonucleotide levels [Dick and Wright, 1985a]. On the other hand, the levels of the second enzyme of the pathway for de novo purine biosynthesis, i.e. glycinamide ribonucleotide synthetase, increases progressively [Hards and Patterson, 1986].

Another metabolic parameter whose evolution seems related to the kinetics of proliferation, is glycolysis. In human embryonic fibroblasts, between population doubling 20 and 40, glucose uptake and lactate

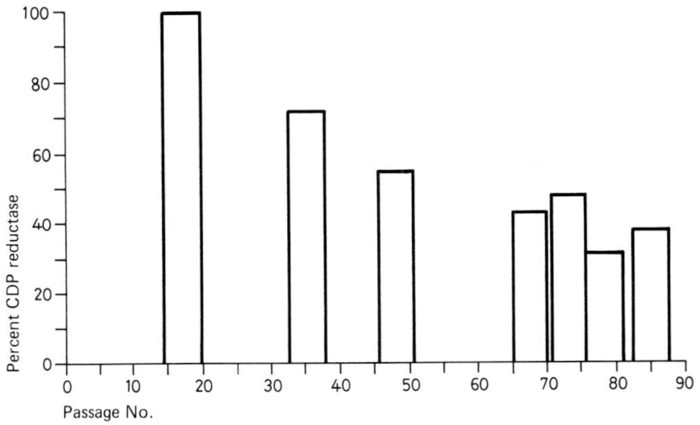

Fig. 41. Relative levels of CDP reductase activity during the life span of fetal lung fibroblasts. CDP reductase activity at passage 17 is 0.230 nmol of deoxycytidine per hour per 3×10^6 cells [Dick and Wright, 1982].

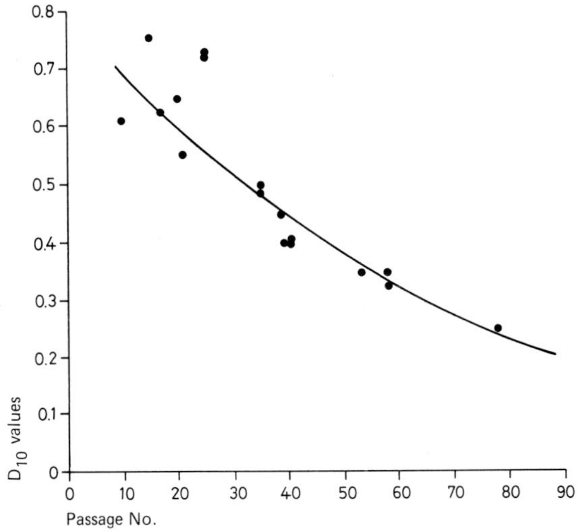

Fig. 42. The resistance of fetal lung fibroblasts, at various passages, to hydroxyurea. Resistance (D 10) is expressed as the concentration of hydroxyurea required to reduce the plating efficiency to 10% [Dick and Wright, 1982].

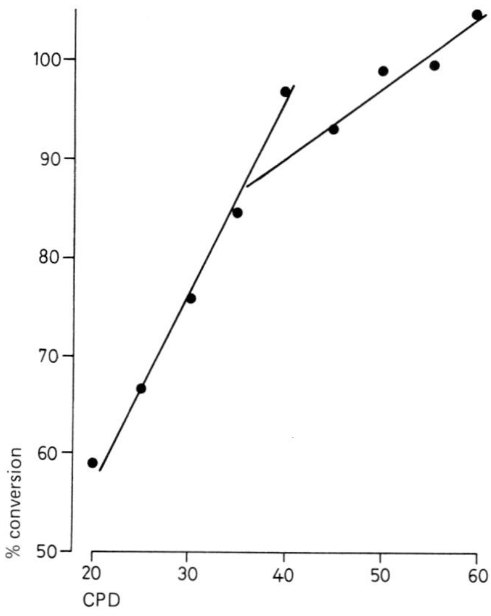

Fig. 43. Mean cellular lactate production at each fifth cell population doubling, expressed as a percentage of the mean glucose uptake [Bittles and Harper, 1984].

production decreased, thereafter, both glucose uptake and lactate production increased [Bittles and Harper, 1984]. It is interesting that the critical change in the biphasic curve of glucose uptake/lactate production corresponds to the transition from phase II to phase III (fig. 1, 26, 43). The same investigators also observed an increase in the specific activity of piruvate kinase, a regulatory enzyme of the glycolytic pathway.

The relative proportion of cyclin, a nuclear protein associated with cell growth, decreased progressively [Celis and Bravo, 1984]. The evolution of the response of a human embryonic fibroblast population to EGF is also interesting [Kaji and Matsuo, 1983]. The magitude of the increase in saturation density due to EGF, progressed during the first half of the population life span and was maximal in middle-passage cells. In the second half of the life span, the responsiveness to EGF decreased (fig. 44). The time when this switch takes place seems also to correspond to the transition between phases II and III.

The activity of several enzymes has been followed during in vitro aging of human cells, and was found either to increase or to decrease [Wang et al., 1970; Milisauskas and Rose, 1973; Srivastava, 1973; Sun et al., 1975; Dell-Orco et al., 1985; Crosti et al., 1985]. A particularly

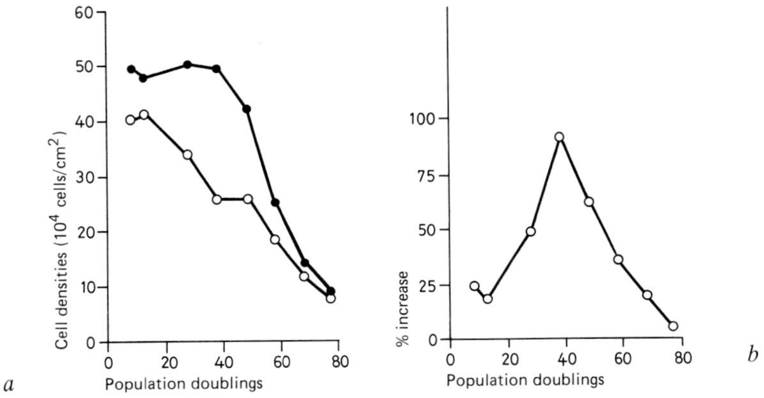

Fig. 44. The effect of EGF on the saturation density of human embryonic lung fibroblasts. *a* variations in the saturation density of EGF-treated (●) and control (○) cells throughout their in vitro life span. *b* Percentage increase in the saturation density of the EGF-treated cells at each population doubling level, based on the saturation density of the control cells at the corresponding doubling level [Kaji and Matsuo, 1983].

interesting finding concerns lactic dehydrogenase and its coenzymes [Minkowitz et al., 1985]. They were assayed at each population doubling and a continuous variation was found, suggesting that the cell culture is in a permanent state of change.

Another interesting enzymatic alteration is the progressive appearance of a new 3′:5′-cyclic AMP-independent histone kinase, raising the possibility that new genes become progressively activated [Kahn et al., 1982].

At the membrane level, progressive modifications have also been reported, some having a pattern with analogies with the kinetics of proliferation. Bosmann et al. [1976] reported a progressive decline in the net negative surface charge and in the activity of a pH 7.8 protease (fig. 45). The V_{max} for saturable sugar transport [Germinario et al., 1980] and the transport of albumin increase [Berumen and Macieira-Coelho, 1977] (fig. 46).

Following the discovery of Goldstein and Brown [1974] of LDL receptors in human skin fibroblasts, and its implications for cholesterol metabolism, several investigators used this approach to analyse cell aging

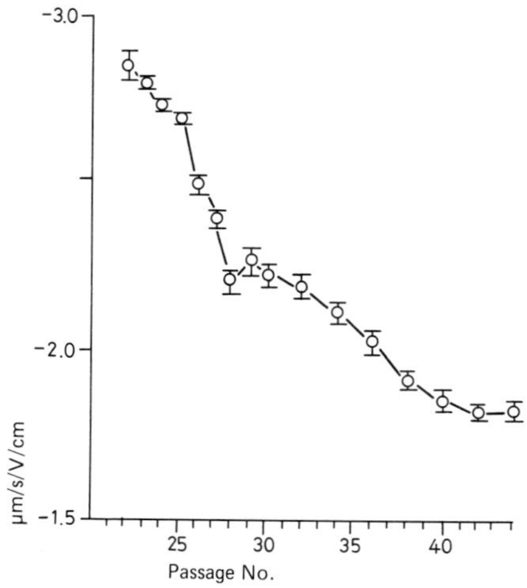

Fig. 45. Electrophoretic mobility of human embryonic lung fibroblasts plotted against passage number of the cells. These data reflect population net negative surface charge [Bosmann et al., 1976].

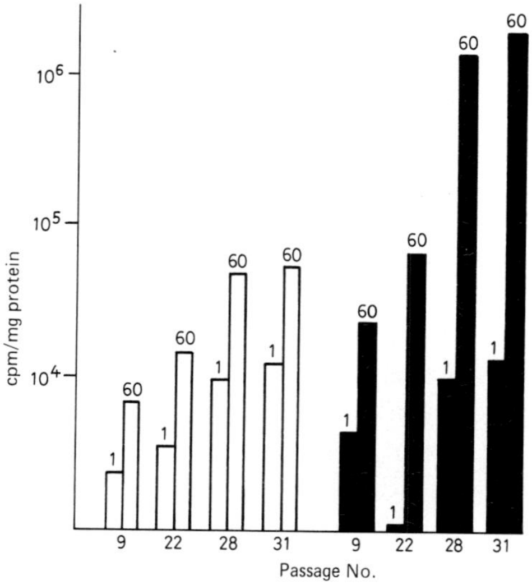

Fig. 46. Incorporation of ^{125}I-labeled seroalbumin at different passages of cultivated human adult lung fibroblasts. The bars indicate 1 min adsorption to the cell membrane and net uptake after incubation for 60 min, in the absence (empty bars) or in the presence (filled bars) of polyornithine [Berumen and Macieira-Coelho, 1977].

in vitro. An interesting finding that has implications for aging and atherosclerosis is the progressive decrease in human fibroblasts aged in vitro, of the binding, uptake and degradation of low density lipoproteins (LDL). The affinity of the LDL receptor remains unchanged, so that the decline must be due to a reduction in the number of receptor sites [Lee et al., 1982]. The aging fibroblasts become also less inducible and less suppressible in the regulation of the surface LDL receptors. The same group obtained similar results with *DL*-3-hydroxy-3-methyl-glutaryl coenzyme A reductase, an enzyme involved in cholesterol synthesis. Pertinent to the problem dealt herein, was the finding that embryonic fibroblasts aged in vitro behaved like fibroblasts from heterozygotes of familial hypercholesterolemia [Lee et al., 1982].

The problem of LDL binding to cells in vitro was also analysed on arterial smooth muscle cells from donors of different ages [Bierman et al., 1979]. The binding did not change with donor age, but LDL degradation decreased in cells from old donors. In the same type of cells, LDL binding and degradation were increased in cells of high population doubling levels [Bierman and Schwartz, 1984]. Acetyl LDL binding and degradation, however, were unchanged. Behrman and Wynn [1984] analysed uptake and degradation of LDL by fibroblastic cultures from normal and atherosclerotic subjects of different ages. Atherosclerosis was found to be significantly correlated with elevated rates of LDL uptake and degradation, and in vivo aging was associated with a significant reduction in cellular LDL metabolism.

A progressive change reported, not with in vitro aging but with in vivo aging, concerns the structure of membranes [Schroeder et al., 1984]. These investigators compared human fibroblasts from donors of different ages and found an increase in the phosphatidylcholine/phosphatidylethanolamine ratio with increasing age of the fibroblast donor. They also obtained, with the use of probe molecules, results suggestive of increased restriction in probe motions with increasing age of the subject.

The variation in the binding of different ligands to the cell membrane expresses also the remodeling that goes on in the cell during serial divisions. Yamamoto et al. [1977] found a decreased concanavelin A (Con A)-mediated cell-cell agglutination. The cellular capacity to bind Con A alone, however, increased when the results were expressed per cell or per cellular protein, but remained unaltered when expressed per unit area of cell surface. After trypsinisation, agglutination occurred to the

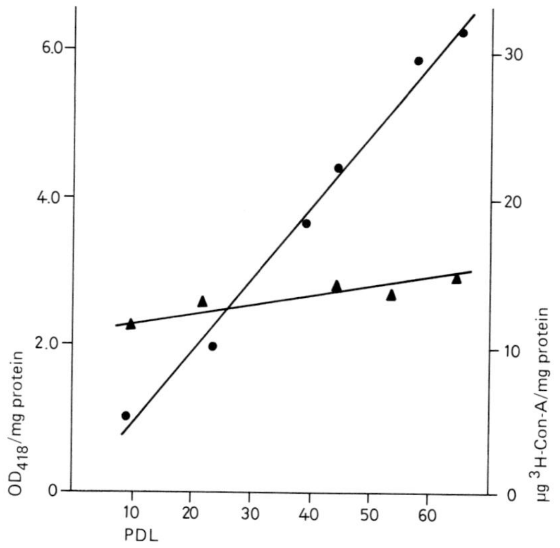

Fig. 47. Adsorption of Con A-coated red blood cells (●) and ^{3}H-Con A binding (▲) to human diploid fibroblasts as a function of population doubling level [Aisawa and Mitsui, 1979].

same extent in both aged and young cells; after hyaluronidase treatment, the agglutination of old cells increased but not that of young cells, a result suggesting that Con A receptors do not change but become masked by other cell components.

Aizawa and Mitsui [1979] developed an interesting method for following the evolution of lectin binding during aging of human fibroblasts in vivo. They absorbed Con A to red blood cells (RBC), then the Con A-coated RBC was adsorbed to monolayers of fibroblasts, and finally the cells were dissolved in 5% SDS and the amount of adsorbed RBC was determined by measuring spectrophotometrically the hemoglobin content. The amount of RBC adsorbed to the fibroblasts via Con A-coated RBC (the RBC coating method) increased continuously throughout senescence in vitro (fig. 47). Binding to Con A alone, however, did not change significantly, as measured after adsorption of (^{3}H)-Con A to the cells. The changes in Con A-mediated RBC binding to a population of fibroblasts are dependent on the percentage of life span completed, are unrelated to cell volume or to the proportion of fast- or slow-dividing cells present, and are independent of metabolic time. However, some rare

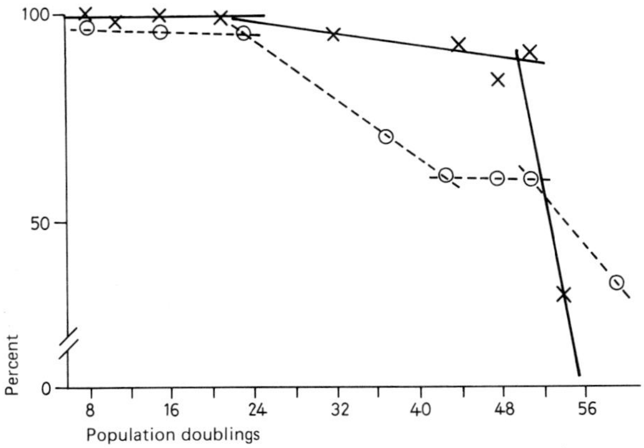

Fig. 48. Maximal percent labeled interphases during a 24-hour period (×) and percent fibrin clot retractin in the presence of cisteamine (○), at different population doublings of human embryonic lung fibroblasts [Azzarone and Macieira-Coelho, unpubl.].

clones present in old cultures show minimal adsorption capacity [Aizawa and Mitsui, 1982]. These observations confirm the finding of Bosmann et al. [1976] that continuous changes of the cell surface start early and progress throughout the population life span. Indeed, both phenomena were correlated by Aizawa et al. Cytochalasin B enhanced adsorption, which suggested that receptor mobility via the cytoskeleton plays a role in the Con A-mediated RBC adsorption.

Germane to this problem of the interaction between the cytoskeleton and the membrane is the finding that cell contractility, measured as a function of the capacity to retract a fibrin clot, becomes more difficult. Furthermore, this decline in the efficiency of cell contraction, follows a pattern identical to that of the maximal number of cells capable of synthesizing DNA during a 24-hour period (fig. 48). Fibrin clot retraction (FCR) [Azzarone and Macieira-Coelho, 1984] was measured at different doubling levels of a human embryonic lung fibroblast population in the presence of 0.5 *M* cisteamine. It was found that the fibroblasts had a steady and maximal FCR during phase II; during phase III they became progressively more sensitive to the inhibitory action of cisteamine; and, finally, during phase IV the retraction of the clot in the presence of the inhibitor was very weak. Contractility of fibroblasts measured by the FCR test, declines in postnatal fibroblasts as compared to embryonic

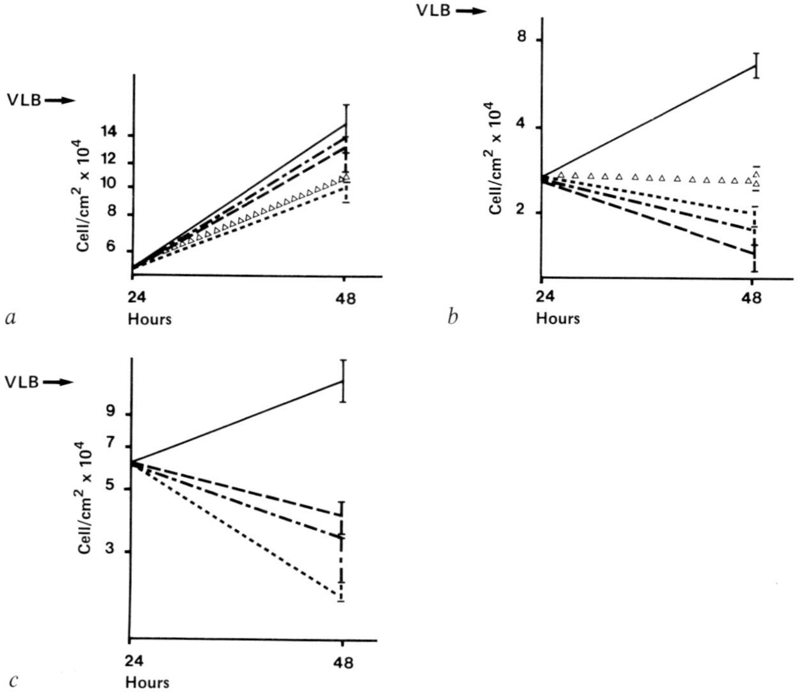

Fig. 49. Growth of human embryonic lung fibroblast at the 48th *a,* 24th *b* and 19th *c* population doublings, in the presence of different concentrations of vinblastine (VLB) controls (—), 0.1 (--), 0.2 (---), 0.4 (----), and 0.8 (△) μg/ml. The bars indicate the extreme values found between 4 samples.

ones, and in skin fibroblasts from patients with certain diseases with disturbances of cytoskeletal elements [Azzarone and Macieira-Coelho, 1984].

The response of serially proliferating human embryonic fibroblasts to drugs used in cancer chemotherapy also varies as illustrated in figures 49 and 50. High population doubling cells were more resistant to the growth inhibitory action of 3 different drugs, than their low population doubling progenitors.

Last but not the least in regard to markers that express the remodeling taking place in dividing cell compartments is the evolution of the response of cultivated human fibroblasts to carcinogens.

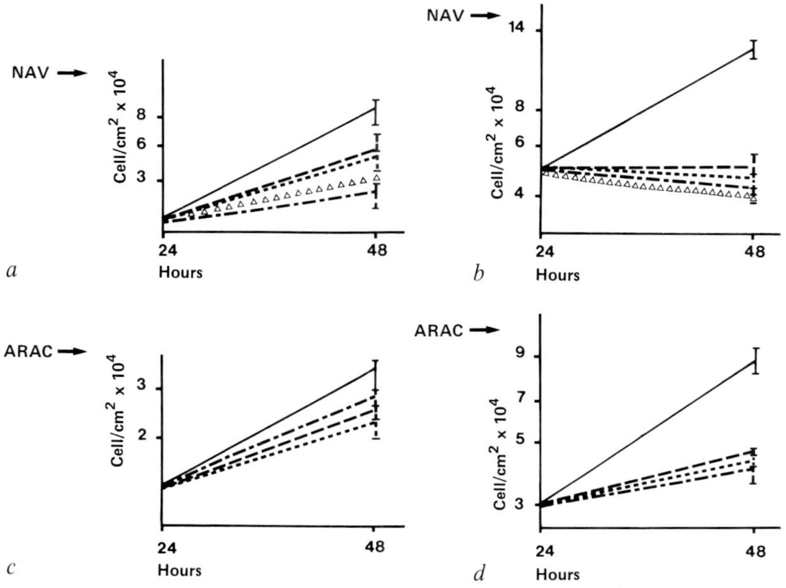

Fig. 50. Growth of human embryonic lung fibroblasts at the 44th (a and c) and 19th (b and d) population doublings, in the presence of different concentrations of navelbine (NAV): 1.0 (--), 2.0 (---), 4.0 (----), 8.0 (△) μg/ml and of aracytine (ARAC): 0.5 (--), 1.0 (---), 2.0 (----) μg/ml. Controls (—).

One that will be mentioned with more detail in the following section, is the response to SV_{40}. Indeed, SV_{40}, like chemical and physical carcinogens, induces in general a partial cell transformation with prolongations of the cell life span when young populations are infected [Pereira-Smith and Smith, 1981]. In terminal cells it induces only DNA synthesis without cell proliferation [Matsumura et al., 1980].

The response of human fibroblasts to ionizing radiation and chemical carcinogens is also representative of the evolution proceeding in this cell system during serial divisions.

With high dose rate ionizing radiation, early passage diploid human fibroblasts are unable to sustain sublethal radiation damage [Cox and Masson, 1974]. Indeed, the survival curves are unshouldered and exponential. However, survival curves of cells that had completed more than 35 generations in vitro are clearly shouldered. Hence, later passage cells gain the ability to sustain sublethal radiation damage. These experiments were first made with embryonic cells and later confirmed with postnatal skin fibroblasts.

The response to low dose rate radiation is different, but is also representative of the changes occurring in the cell population.

When the radiation was applied early during the cell life span, it either had no effect or extended the cell population life span [Macieira-Coelho et al., 1977; Croute et al., 1986]. As the cell age in vitro and with development in vivo, they become more sensitive to radiation which shortens their life span. This increased sensitivity to low dose rate ionizing radiation seemed to be associated with an increased fragility of chromatin [Icard et al., 1979]. It was these results obtained with low dose rate ionizing radiation that led to the hypothesis described above, that the remodeling going on in dividing cells is the result of the genome reorganization taking place during cell division [Macieira-Coelho, 1979, 1980, 1981]. In addition, these experiments buttressed the view, that mutations are not responsible for the aging of dividing cells. Other dose rates of ionizing radiation were found to shorten the survival of cells, with increasing age in vitro [Ban et al., 1980b].

The follow-up of the response of human diploid fibroblasts to chemical carcinogens has shown that low passage cells respond to carcinogen and mutagen treatment by the acquisition of some markers of a transformed phenotype, while higher passage cells remain refractory to the insult [Milo and Casto, 1986]. One may wonder if this evolution of the susceptibility of fibroblast populations to chemical carcinogens is related to the finding that early passage cells are susceptible to the inhibition of queuine uptake by a tumor promotor [Elliot and Katze, 1986]. Late passage cells, however, become refractile to this induced inhibition. Queuine is a nucleoside found in the first position of the anticodon for some tRNAs. This property of early passage cells seems to be due to a factor secreted by the young fibroblasts. Furthermore, the growth factor stimulation of queuine uptake is enhanced by the tumor promotor in late passage cells but inhibited in early passage cells. The kinase activities of growth factor receptors and phorbol ester-stimulated protein kinase C suggest that phosphorylation is a crucial control event in the uptake of queuine [Elliot and Katze, 1986].

Testing the error/mutation hypothesis of cellular senescence, also yielded results suggestive of genetic changes progressing through serial divisions [Gupta, 1980]. The mutation frequencies for diphteria toxin and thioguanine resistance was found to follow a slow, linear, time-dependent increase during the first $^2/_3$ of the life span of a human fetal lung fibroblast population, which is consistent with the random occurrence of mutations. It was also found that treatment with two mutagens (ethyl methane sulfonate and methyl-N′-nitro-nitroguanidine) did not have a life-shortening effect on the cells. These experiments and those reported above, obtained by Macieira-Coelho et al. with low dose rate ionizing radiation, argue strongly against a causality role of mutations in cellular senescence.

This is a long list of events demonstrating how cells can change and evolve through serial divisions. In the case of fibroblasts, are these progressive changes the result of a programed developmental process, due to successive shifts to definite cell types [Martin et al., 1974; Mollenhauer and Bayreuther, 1986], or is it a random evolution with the appearance of a multitude of cell types?

Some results could suggest the first hypothesis, such as those obtained by Mollenhauer and Bayreuther [1986], showing the successive development of definite morphologic cell types. The decline in 6-keto-prostaglandin-F_1-alpha-producing activity with increased production of thromboxane B_2, prostaglandin-F_2-alpha- and prostaglandin E_2-like metabolites [Murota et al., 1979] could also favor the first hypothesis.

It is possible that the situation varies with the cell system. Postnatal skin fibroblasts may follow through serial divisions in vitro, a developmental process with quite definite cell types, due to their stage of differentiation in the respective tissue. Other types of postnatal cells may follow one or the other evolution. Embryonic fibroblasts, due to the multipotentialities of the initial population, may follow a more random evolution. This problem will be discussed further in the follow section, dealing with the terminal cell where the probability of engaging into the cell cycle has become very low or eventually nil. The problem of seeing the evolution of some dividing cells compartments in terms of a programed or a random event is for the moment speculative. But the immense amount of work available advanced our understanding of the mechanisms involved in this progressive loss of the probability of dividing and of the changes occurring at the molecular and supramolecular level through proliferation.

A model has been proposed that could provide a broad explanation for the events taking place [Macieira-Coelho, 1983; Macieira-Coelho et al., 1987]. As described in chapter III.1 and in the beginning of this section, division is the inevitable event that modifies a cell, through reorganization of the genome. This can lead to a discoordination in the synthesis

Gene reorganization during cell division

↓

Changes in the cellular scaffold

↓

Changes in cell attachment, spreading and the anchorage of DNA

↓

Decreased mobility of the cytoplasm and the nuclear cage

↓

Reorganization in the high order structure of DNA

↓

Decline of the probability of initiating the division cycle

Fig. 51. Schematic representation of the putative chain of events leading to an increased cell volume and decreased motility, and feeding back on the genome during the growth decline of serially proliferating cells.

of different molecules, inter alia those responsible for the organization of the cell scaffold. Indeed, some of the genes responsible for the synthesis of cytoskeletal elements seem to be involved in this reorganization [Icard-Lipkalns et al., 1986]. As described in chapter VI.2, there is evidence favoring the view that cell movements are necessary for the initiation of the division cycle and of DNA synthesis because of the topological changes they create at the molecular and supramolecular level. With the discoordination of the synthesis of the components of the cellular scaffold, the cell increases in size and loses progressively its motility and contractility, so that the flow of movements from the periphery to the nucleus declines. Cell size in itself must create profound changes through modification of conformation and of the association between molecules. Since DNA is anchored on a nuclear matrix which is linked to the cell membrane through the cytoskeleton, its high order organization is influenced by the dynamics of the cell scaffold and by cell attachment to the extracellular matrix. Thus, the changes in these cellular and extracellular structures will feedback on the genome and accentuate its reorganization. This succession of events is schematically represented in figure 51.

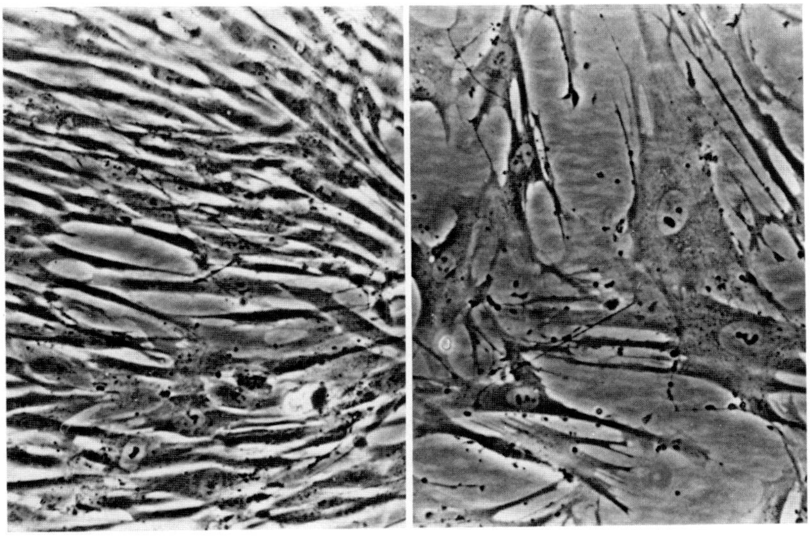

Fig. 52. Phase II (left) and phase IV (right) human embryonic lung fibroblasts serially propagated in vitro.

Obviously, this is a broad scheme representing the main events leading to a postmitotic cell. It can be easily seen how the other mechanisms that are involved in the growth decline can be grafted into this schematic view. We will now review what is known about the terminal cell originated in vitro.

2. *The Terminal Stage (Phase IV): Degeneration or Differentiation?*

Besides the striking changes in the kinetics of cell proliferation, mentioned above, taking place during the last 3–4 doublings, a multitude of precipitating events support the idea of a final distinct phase. Most of these events have been observed in fibroblasts.

The volume of the cell increases abruptly [Bemiller and Miller, 1979] (fig. 52), probably due to sudden modifications in the organization of cytoskeletal elements; in particular those at the level of microtubules [Van Gansen et al., 1984; Raes et al., 1984] which must be responsible for the prolongation of cytokinesis described by Sisken and Bonner [1979] in the

Table III. Percent nuclei with altered nucleoli and chromatin found at different PDL during the serial subcultivation of the human embryonic lung fibroblast line ICIG-7

PDL	%
10th to 40th	0
41st to 49th	5
50th to 56th	98

terminal stage and to the abnormal mitoses observed by several investigators (see chapter VI.2.d.). This profound reorganization of the microtubules may also be the reason why it is possible to distinguish phase IV cells with an antibody against microtubule-associated proteins [Ohno et al., 1986]. The reorganization of the cellular scaffold is also probably implicated in the decondensation of chromatin seen with conventional fixation [Brock and Hay, 1971; Puvion-Dutilleul and Macieira-Coelho, 1982] and for the altered distribution and anchorage to the lamina densa of the 30-nm chromatin fibers (fig. 24). On the other hand, the chromatin condensation described during terminal differentiation of chicken lens [Sanwal et al., 1986] corresponds probably to apoptosis and is thus a prelude to cell death.

The frequency of nuclei with altered chromatin fibers at different doubling levels of a human embryonic lung fibroblast population aged in vitro, was determined by direct screening of electron microscope preparations (table III). It was found that up to the 41st population doubling, altered nuclei could not be detected; from the 41st to the 49th doubling, 5% of the nuclei had a modified organization of chromatin and from the 50th doubling on up to the end of the cell population life span (56th doubling) the percent of altered nuclei increased abruptly from 5 to 98% [Puvion-Dutilleul et al., 1984]. This shows that the rapid fall in the fraction of cells synthesizing DNA per unit of time is accompanied by dramatic changes in the nucleus simultaneously with the increase in cell size.

Identical measurements were made in early passage cells obtained from donors of different ages. It is interesting that the percent of nuclei with gross changes, visible by direct observation, was increased in postnatal fibroblasts as compared to foetal ones, was increased in cells from a 96-year-old donor as compared to middle age donors, and in cells from a patient with Werner's syndrome as compared to age matched normal

Table IV. Percent nuclei with altered nucleoli and chromatin found in low PDL cultures of human skin fibroblasts

Cell line	Age, years	PDL/maximal number of doublings	Nucleoli	Chromatin
Fetal				
ICIG-9		4/50		
AG 4525		6/50		
Postnatal				
MVL	30	12/30		17
AG 4353	59	6/40		7
AG 4059	96	5/29	63	63
AG 780	60	6/19	89	89
Werner's syndrome				

donors (table IV) [Puvion-Dutilleul and Macieira-Coelho, 1983]. In addition, the cells from the 96-year-old normal donor and from the Werner's syndrome patient had altered nucleoli. Thus, it is possible that in very old age and in pathological aging, the number of mesenchymal cells with phase IV type chromatin, increases. Indeed, we also found an increase of altered cells in a 30-year-old donor with diabetes and advanced atherosclerosis [unpubl. results].

The spacing between fibers at the periphery of nuclei of human embryonic fibroblasts was measured with the same image analysis method described above (fig. 24). It was found that, contrary to the density which declines progressively with a pattern similar to the prolongation of the initiation of DNA synthesis (see chapter VI.2.b.), the spacing between chromatin fibers increased abruptly at the end during phase IV, coinciding with the fall in the number of cells synthesizing DNA during a 24-hour period (fig. 25).

This sort of measurement was also made on fibroblasts cultivated from skin biopsies of normal donors of different ages and of patients with Werner's syndrome (fig. 53). It was found that the number of cells with increased spacing between fibers increased with the age of the donor. Cultures obtained from skin biopsies from Werner's syndrome patients had a larger fraction of cells with an increased spacing as compared with

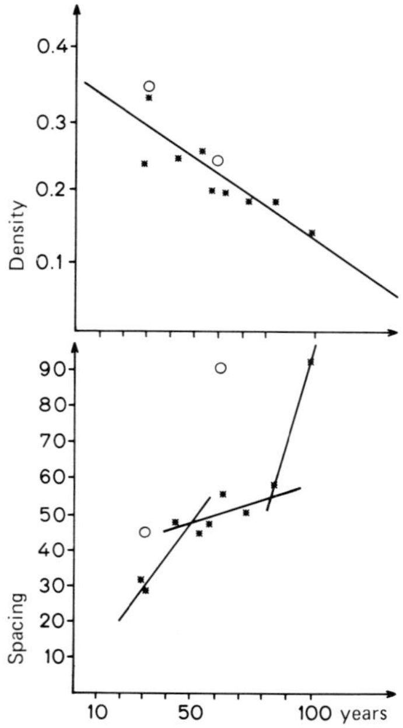

Fig. 53. Mean density and spacing of the 30-nm chromatin fibers found in early passage cultures of skin fibroblasts from normal donors (★) and from Werner's syndrome patients (○) of different ages.

age-matched controls, and the difference increased with the age of the patient.

The changes in the arrangement of chromatin fibers are detected only in cells that migrate from the explants prepared from skin biopsies. In the biopsy itself one cannot see the modifications [Macieira-Coelho and Puvion-Dutilleul, unpubl. observations]. This suggests that the cells have to attach and migrate on a substratum for the increased spacing between the fibers to become apparent; it reinforces the idea that the chromatin changes depend at least in part from modifications of the scaffold on which they are anchored, and that migration and stretching of the cells reveal the latent alterations that accumulate with senescence in chromatin. This is germane to the finding that mouse fibroblasts in collagen gels seemed stable and did not display the characteristics of an older phenotype [Lerberghe and Van Gansen, 1986], despite the fact that their actin patterns are different when cultivated on glass. When the fibroblasts migrated and left the gel, they returned to the old phenotype.

The chromatin from phase IV cells also has an increased thermolability in an alkaline medium [Puvion-Dutilleul et al., 1984] and a decreased template activity [Hill et al., 1978]. After Miller spreads the nucleosomal structure of the 10-nm chromatin fibers disappears to a great extent in phase IV cells aged in vitro and in cells from patients with Werner's syndrome [Puvion-Dutilleul and Macieira-Coelho, 1983] (fig. 54). Since analysis by biochemical methods did not reveal a loss of nucleosomal structure [Dell'Orco et al., 1986] the changes observed with Miller spreads are probably due to shear forces on a more fragile structure, resulting from the ionic detergent, the alkaline pH and the hypotonic solution used during the preparation for electron microscopy. The fragility of chromatin could be due to changes in conformation accompanying the dramatic increase in cell size; the disappearance of the nucleosomal structure after Miller spreads may be secondary to alterations of the synthesis of histone H1 [Mitsui et al., 1980].

Modifications of the physicochemical properties of chromatin from phase IV are also revealed by the different circular dichroism spectra under certain conditions of extraction [Maizel et al., 1975]. They might be also due to nonhistone chromosomal proteins which show differences in gel electrophoretic profiles.

There is also an increase in the component P8 of chromatin binding proteins [Stein, 1975]. The increase in extractable nucleoproteins [Hamelin et al., 1984] seems to be due mainly to the accumulation of residual acidic proteins [Mitsui et al., 1980]. In chicken fibroblasts, no changes in circular dichroism were observed, but there were differences in the incorporation of labeled aminoacids into histone and non-histone proteins [Courtois, 1974]. Terminal postmitotic cells lose the ability to synthesize detectable amounts of four major nuclear proteins that are found in young cells [Sakagami et al., 1979]; on the other hand, old cells synthesize two new major proteins that are undetectable in young cells.

Other changes in DNA during phase IV concern DNA repair which was discussed in chapter III.2.b., and seems related to the difficulty in initiating DNA synthesis. During phase IV there is also an increase in small molecular weight DNA, detected in alkaline sucrose gradients [Puvion-Dutilleul et al., 1984] (fig. 55), which was later shown to have a circular form [Kunisada et al., 1985] (fig. 56). Circular and linear small molecular weight DNA is also found in cells cultivated from Werner

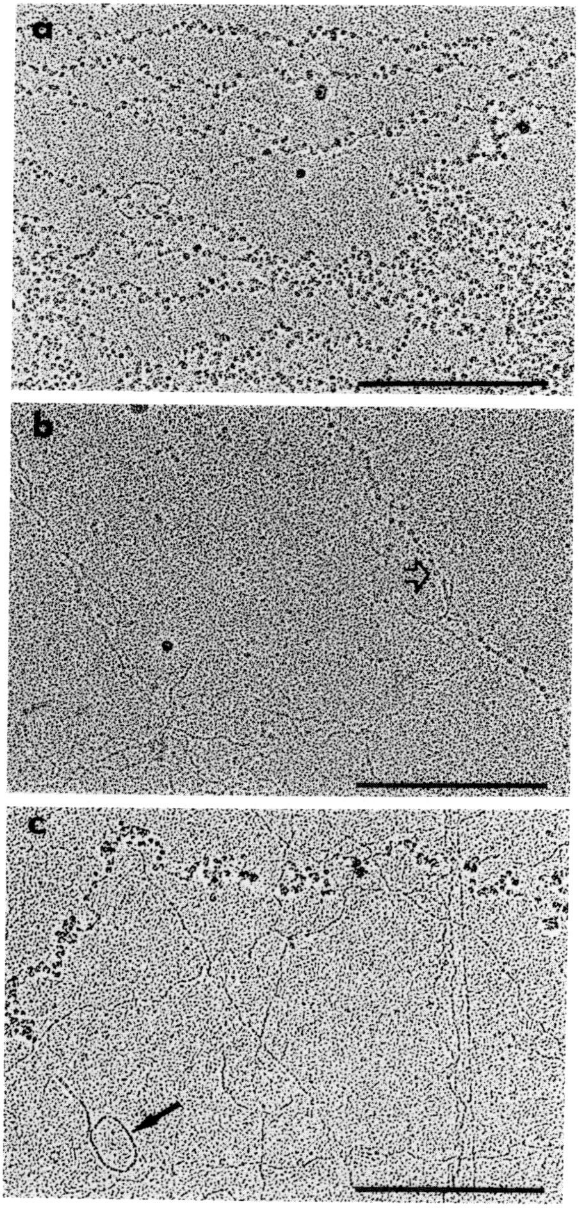

Fig. 54. Ten-nanometer chromatin fibers from phase II human embryonic lung fibroblasts *a,* from early passage skin fibroblasts from a 59-year-old normal donor *b,* and from a 60-year-old patient with Werner's syndrome *c* [Puvion-Dutilleul and Macieira-Coelho, 1983].

patients [Puvion-Dutilleul and Macieira-Coelho, 1983] (fig. 54) albeit not amplified when compared with cells from age-matched normal donors [Kunisada et al., 1985]. The extrachromosomal circular DNA was found to contain sequences of the interspersed highly repetitive Kan I and Alu families [Riabowol et al., 1985], to have sequences homologous to the alpha-globin gene, that is not expressed in fibroblasts, and to a multigene family that is expressed, the beta-actin gene [Icard-Liepkalns et al., 1986]. This shows that the circles are heterogeneous and originate from different parts of the genome.

Concomitant with the presence of extrachromosomal DNA, there is a loss in the hybridization with genomic DNA, of the probes for a specific family of tandemly repeated sequences [Shmookler Reis and Goldstein, 1980] and of the probes for the alpha globin and beta actin genes [Icard-Liepkalns et al., 1986]. Amplification of extrachromosomal small circular DNA increases during in vivo aging [Kunisada et al., 1985] and was also found to be increased in a murine model of accelerated senescence [Yamagishi et al., 1985]. In humans, the amplification of restricted size class of circular DNAs seems to occur only in the later part of the human life span [Kunisada et al., 1985].

The uneven distribution of DNA between sister cells, which occurs at a constant rate through out the life span of the fibroblast stem cell compartment, increases abruptly during the last 2–3 doublings [Macieira-Coelho et al., 1982; Macieira-Coelho and Puvion-Dutilleul, 1985] (fig. 20) and one might wonder if it is related with the amplification of the small circular extrachromosomal DNA.

Another change occurring in chromatin, is an abrupt increase in the proteolysis of H1 histone [Houde, 1986]. It may be part of the general increased proteolytic susceptibility of proteins observed after the last doubling by Bradley et al. [1975].

Only infection with DNA viruses such as human cytomegalovirus and SV40 can induce DNA synthesis in the terminal cell [Ide et al., 1983, 1984a; Gorman and Cristofalo, 1985]. Infection with SV40 virus, however, cannot induce cell proliferation [Matsumura et al., 1980]. The more rapid induction of proliferation and immortalisation of terminal cells by SV40 alone, previously reported [Girardi et al., 1965], must be a rare event. An efficient immortalization though, can be achieved, if the cells are transfected both with the SV40 early region encoding the T antigen and with the hypoxantine-guanine phosphoribosyl transferase gene [Mayne et al., 1986]. It is interesting that to achieve immortalization the

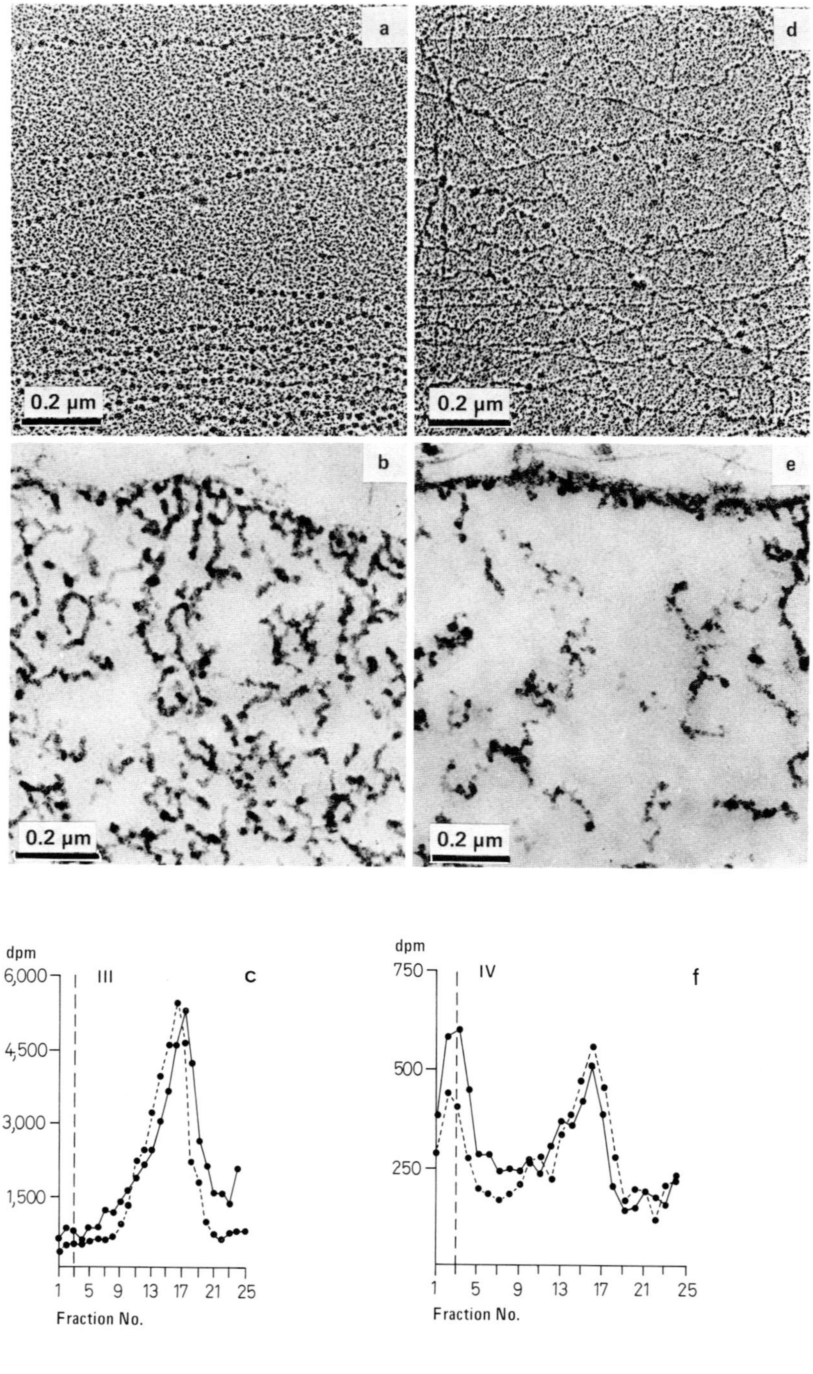
a
d
0.2 µm
0.2 µm
b
e
0.2 µm
0.2 µm
dpm
6,000
4,500
3,000
1,500
III
c
1 5 9 13 17 21 25
Fraction No.
dpm
750
500
250
IV
f
1 5 9 13 17 21 25
Fraction No.

transfected cells have to go through phase IV. This buttresses the idea of some fundamental change occurring in the terminal cell, which complements the changes induced by the virus and helps to modify the genetic information allowing unlimited divisions.

It was suggested that phase IV SV_{40}-transformed, but not immortalised human fibroblasts, contrary to normal cells, lose the ability to enter a visible G1-arrested state when mitogen deprived [Stein, 1985]. Concomitantly the rate of cell death increases steadily. Both normal and transformed cells, however, lose the ability to respond to mitogens. Hence, it was hypothesized that the transformed cells undergo the same aging process as the normal fibroblasts, but they end their life span in crisis because they continue their cell cycle traverse coupled with cell death [Stein, 1985].

The enzyme 5′-nucleotidase, which increases in normal phase IV cells, was found to be at a lower level in SV_{40} immortalized human fibroblasts [Sun et al., 1979].

The experiments of Ide et al. [1984b] led to the conclusion that the clock of aging, progresses in transformed cells as in normal ones. These investigators used a temperature-sensitive mutant virus to infect the cells. When cultured at 34 °C, the life span of the cells was prolonged. When the transformed cells whose life span had been prolonged were shifted to 39.5 °C, they immediately reached the end of their proliferative life span. A similar event was observed during the life span prolongation of human fibroblasts, with hydrocortisone; in cells grown beyond the number of doublings of the controls, removal of the hormone led to the immediate end of proliferation [Macieira-Coelho and Loria, 1974, 1976].

It was postulated that after stimulation with growth factors, the terminal cell procedes through G1 but stops just before the initiation of DNA synthesis. This was concluded from results showing that it is possible to induce the expression of 11 cell-cycle-dependent genes, including

---

Fig. 55. Phase III *a–c* and phase IV *d–f* fibroblasts prepared in parallel for electron mircroscopy and alkaline sucrose gradient centrifugation. *c, f* Sucrose gradient centriguation. *b, e* 30 nm chromatin fibers. *a, d* Miller's spreads. The curves represent the radioactivity found in the different fractions collected from the gradients. — = Cells treated the same way as for the loosening procedure; ---- = cells prepared by conventional methods. The vertical dashed lines indicate the fraction where lambda phage was found [Puvion-Dutilleul et al., 1984].

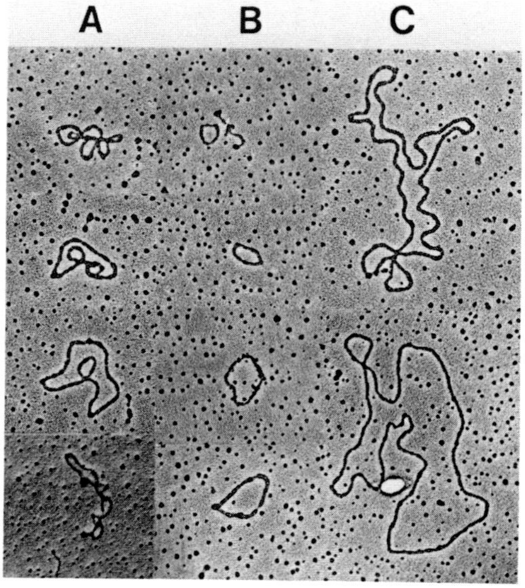

Fig. 56. Electron micrograph of circular DNA from human embryonic fibroblasts in phase IV. DNA preparations of 4.91 ±1.1 *A*, 1.5 ± 0.4 *B* and 17.2 ± 2.4 kb *C*. The latter correspond to mitochondrial DNA. Length measurements were made relative to the plasmid pBR 322 with 4361 bp. ×51,000 [Icard-Lipkalns et al., 1986].

the two genes for thymidine kinase and histone H3 that usually are maximally expressed at the G1/S boundary [Rittling et al., 1986]. There are, though, differences in the expression of those cell-cycle-dependent genes; while there is an amplification and overexpression of the C-Ha-ras gene in phase IV [Srivastava et al., 1985], the expression of the c-myc gene is decreased [Dean et al., 1986]. Other authors, however, found increased levels of c-myc RNA [Kihara et al., 1986a; Paulsson et al., 1986] as well as of c-fos. Microinjection of c-H-ras induces DNA synthesis in reversible quiescent cells, but not in phase IV cells [Lumpkin et al., 1986b]. This is further proof that there are fundamental differences between these two states. Nuclear fluorescence after quinacrine dihydrochloride staining also favors the idea that phase IV-stimulated cells proceed through G1 and stop at the G1/S boundary [Gorman and Cristofalo, 1986]. It has also been claimed that the synthesis of thymidine triphosphate, an event that normally begins in late G1, is inducible in the terminal cell [Olashaw et al., 1983]. This observation could be due to the fact that these authors used preterminal cells since Kihara et al. [1986a] could not find an uptake of ^{3}H-thymidine into the acid-soluble pool of true postmitotic cells stimulated by fresh serum.

Other events occurring in postmitotic cells stimulated by growth factors are the actin fibers reorganization [Paulsson et al., 1986], and the uptake of 2-deoxyglucose and uridine [Kihara et al., 1986a]. RNA synthesis and polyamine accumulation, however, do not take place. Since there are many events that are not induced, the idea that cells complete the G1 period may not fit the multiple pathways that occur during the preparation for DNA synthesis [Kihara et al., 1986a]. Indeed, the phosphorylation of ribosomal S6 proteins, which is a prerequisite for the activation of protein synthesis during G1, was not enhanced in stimulated phase IV cells [Kihara et al., 1986b]. However, cell-free extracts from the latter could stimulate the S6-kinase activity. The authors suggested that a regulatory molecule might be lost during cell lysis.

Another event that does not occur after stimulation of phase IV cells is the expression of the nuclear antigen K-67, which, in proliferating cells, is induced in the S period [Paulsson et al., 1986].

The activity of DNA polymerase alpha is normal on a per cell basis, but becomes significantly lower when normalized to total protein content, because of dilution due to the sudden cell enlargement [Pendergrass et al., 1985]. This shows how changes in intracellular topology can have a profound influence on cell metabolism through the dilution of molecules and changes in their respective affinities, and contribute this way to the postmitotic state.

DNA polymerase-beta declines also at the very end [Krauss and Linn, 1982], which would be consistent with the reports that DNA excision repair becomes less efficient in phase IV [Painter et al., 1973; Hart and Setlow, 1976].

Abrupt changes in other enzymes have also been reported. Thus, Sun et al. [1979] and Raes et al. [1981] observed an increased 5′-nucleotidase activity. The latter enzyme hydrolyzes all deoxyribo- and ribonucleoside monophosphates; it may, through a depletion of the nucleoside phosphate pools, contribute to the difficulty in initiating DNA synthesis in the terminal cell. Gamma-glutamyltransferase, an enzyme of the gamma-glutamyl cycle, which may be involved in amino acid transport, increases [Takahashi et al., 1978]. It could be a reflection of the changes in the membranes where the transferase is located.

Transglutaminase activity also increases during phase IV [Dell'Orco et al., 1985]. It is interesting that with this enzyme, the significant decreased activity that occurs during phases II and III, is reversed in the terminal stage. The same type of reversion occurs with collagen-galactosyltransferase and collagen-glucosyltransferase [Carnicero et al., 1981]. Although the reason for this reversion in the specific enzyme activity is unknown, it favors the idea that new regulations develop during the last division. This is also suggested from the increased production of prostaglandin F_2alpha and E_2 after addition of adrenalin [Mets et al., 1979].

The synthesis of proteins in phase IV cells has also been approached with different methodologies. In human skin fibroblasts, the rate of protein synthesis/cell accelerated nearly fourfold; however, when the rate was expressed as a function of protein concentration no difference was observed [Shevitz et al., 1986]. An increased utilisation of aminoacids from putrescine through the catabolic pathway has also been reported [Cheng and Chan, 1986]. On the other hand, Pool [1982] separated on SDS polyacrylamide gels, radioactive labeled protein extracts obtained from young cells arrested in low serum medium and from terminal cells selected by bromodeoxyuridine. He could this way detect quantitative differences in the labeling of different polypeptides. In addition, terminal cells incorporated label into a 33.1-kdalton polypeptide that was not present in the extracts from young cells. Engelhardt et al. [1979] separated protein extracts from mass cultures of young and terminal cells, with 2D electrophoresis and observed that among more than 500 peptides, two were significantly increased in the latter cells. Lincoln et al. [1984] used the same separation technique with protein extracts from cloned populations and found that terminal cells synthesized two polypeptides not detected in the younger cells. One of these polypeptides could eventually be detected in young cells continually maintained in a growth-arrested state. These authors concluded that although the evidence favors the idea that these two polypeptides are newly synthesized proteins, the possibility that they arise from post-translational modifications, must still be considered. On the contrary, van Keuren et al. [1983] could not see new proteins or satelite spots in two-dimensional electrophoresis in late passage skin fibroblasts from a normal subject and from a patient with trisomy 21. Four proteins, though, were significantly altered in density in both cell populations. Ramaekers et al. [1984] on the other hand, detected a 55-kdalton protein doublet differing from tubulin, during terminal differentiation of lens cells in vitro.

Several events associated with the membrane, are characteristic of the terminal stage. There is a larger increase in AMP level after addition of adrenalin and a smaller increase after addition of prostaglandin E1 [Haslam and Goldstein, 1974; Polgar et al., 1978]. The level of the cAMP dependent protein kinase increases 2 to 3-fold when expressed per milligram of protein and 5 to 7-fold when expressed per cell [Liu et al., 1986]. A membrane associated enzyme, gamma-glutamyl transpeptidase, also increases 13-fold when human lung embryonic cells enter phase IV [Takahashi and Zeydel, 1982].

The synthesis of glycosaminoglycans decreases [Schachtschabel and Wever, 1978] but there is a relative increase in the synthesis of heparan sulfate [Wever et al., 1980].

Under certain experimental conditions the autophosphorylation of the EGF receptor after addition of the ligand does not take place [Carlin et al., 1983a, b], the structure and function, however, of the receptor seems to be normal [Chua et al., 1986]. Paulsson et al. [1986] could not find any difference in the number of PDGF receptors of phase IV cells. The receptors were also functional with regard to PDGF-induced autophosphorylation. In these experiments an antiserum against phosphotyrosine was used, which precipitates only those receptors which are phosphorylated on tyrosine residues. Hence, it seems that there are no gross defects of the phosphorylation of growth factor receptors in the postmitotic cells.

Several experiments suggest modifications of other receptors during phase IV. Thus, there is a significant increase in the adsorption of red blood cells to concanavalin-coated fibroblasts [Aizawa and Kurimoto, 1979a, b]. Scatchard plots obtained with another lectin *(Ricinus communis),* which is specific for beta-*D*-galactosyl residues, showed that in the final stage of the embryonic liver fibroblast stem cell compartment, there is a switch from two classes to one class of binding sites. Adult liver cells, though, are monophasic from the beginning of their life span in vitro [Aubery et al., 1980]. This suggests that adult liver mesenchymal cells, either have a large component with some characteristic of phase IV cells, or that although still capable of dividing have some markers present only in embryonic cells which have exhausted their division potential. It is possible also that this switch to one class of binding sites is characteristic of a GO state rather than of a postmitotic state.

The transport of molecules is also affected. The K_m for 2-deoxy-*D*-glucose transport did not vary during senescence of human postnatal skin fibroblasts [Germinario et al., 1980] but increased in postmitotic embryonic lung fibroblasts [Cremer et al., 1981]. The values for the Vi_{max} of 2-deoxy-*D*-glucose transport per cell increased progressively with the percentage of life span completed and jumped to high levels during phase IV [Germinario et al., 1980]. A similar pattern takes place in the response to polyornithine stimulation of albumin transport [Berumen and Macieira-Coelho, 1977] (fig. 46).

Uptake of zinc, on the other hand, which is stable during most of the fibroblast stem cell compartment life span in vitro, decreases during phase IV [Sugarman and Munro, 1980]. At this stage, the amount of zinc per cell does not vary, but is significantly reduced if expressed per microgram of cell protein. This means that zinc is diluted in the terminal cell and again (see above), it shows how the change in cell volume can by itself interfere in the cell metabolism, in this case, changing the intracellular concentration of a metal that acts as cofactor in many enzyme reactions.

Some interspecies differences seem to exist, which is not surprising. Terminal chinese hamster fibroblasts, for instance, utilize more glucose [Ryan and Pace, 1975] while human cells do not [Cristofalo and Kritchevsky, 1965].

The increased accumulation of lysosomal-associated structures cannot be considered characteristic of phase IV. They must be secondary to the nondividing state since they also accumulate in young cells maintained for long periods of time under conditions of contact inhibition of growth [Macieira-Coelho et al., 1971; Brunk, 1973; Brunk et al., 1973; Collins and Brunk, 1976].

The list of events taking place during the last 3–4 divisions of fibroblasts is quite impressive and it seems obvious that the cells reach a final distinct state.

In regard to the problem of cell proliferation, thanks to the considerable amount of research, one can now understand why there is an increasing difficulty in initiating the division cycle and in completing it. As described later, the mechanisms leading to the slow response to growth stimuli are multiple, as should be expected, and are the result of cumulative events which lead to a dramatic modification of the cell after serial divisions. But in the case of the fibroblast, is this a degenerative process or a developmental one? Are the cumulative steps preceding the final stage the important ones in the physiology of aging?, and is the last step implicated only in pathology or is it also important in the remodeling taking place through the animal life span?

As usual, the answer is probably in between. It is obvious that in old age our cells are still capable of plenty of divisions, and what seems to take place is a deregulation in the proliferative response rather than the absence of the capacity to divide [see, as an example, Roth and Adelman, 1974]. Hence, in this respect it seems that the important feature for the physiology of aging are the small, subtle, cumulative events created through divisions long before the cells reach the last step [Macieira-Coelho and Azzarone, 1982]. But it is also possible that serial divisions originate cells with different properties, which are called to play a role in the physiology of the organism. With aging there could be more or less pronounced shifts in these cell populations, which would constitute part of the aging syndrome. In pathological situations, these shifts could be more pronounced at the local level; an example of such a situation would be the atherosclerotic plaque (see chapter II.1).

Cristofalo [1972] and Martin et al. [1974] considered the possibility that the serial divisions of fibroblast-like cells constitute a differentiation process. They proposed that the fibroblast stem cell compartment undergoes a sort of terminal differentiation, analogous to that observed with other stem cells in vivo, such as hematopoietic cells or myoblasts. Based on the striking changes in morphology, Martin et al. [1974] suggested a two-stage model of clonal senescence: stage I progressing via a process of clonal attenuation and leading to a terminally differentiated cell, and stage II would involve the senescence of these postreplicative cells.

This concept led to the attempt to reactivate DNA synthesis in these putative terminal differentiated cells, in heterokaryons with actively replicating cells, as has been previously done with macrophages, lymphocytes and erythrocytes [Harris et al., 1966, 1969]. Heterokaryons with some immortal cell lines could rescued the old nucleus [Norwood et al., 1975]. These experiments are described in chapter III.2a. Although these experiments do not prove the terminal differentiation hypothesis, they make an error catastrophe mechanism unlikely.

Other authors adhered and tried to explore further the terminal differentiation hypothesis. Bell et al. [1978] are among them, but followed a different course suggesting that the departure of cells from the cycle is not an aging process. They also postulated that cells are not required to pass through a fixed number of divisions before leaving the cycle permanently, since at any PDL there is a high heterogeneity of interdivision times and of clone sizes. This latter assumption remains speculative since it is impossible to know with certainty, the history of a noncycling cell.

The same authors proposed that in the organism there are three types of fibroblasts: potentially cycling (PCF), cycling (CF) and further differentiated fibroblasts (FDF). The interrelationship among the three classes was schematically represented as follows:

PCF $\leftrightarrows$ CF $\rightleftarrows$ FDF.

In other words, according to Bell et al. [1979], the departure from the cycle would be a physiologic process, so that terminal nondividing cells would be induced into this state at any age, to play a role in the tissues yet to be determined. Bell and co-workers claimed that they can induce this state by plating the cells in vitro into collagen matrices [Sarber et al., 1981], a situation analogous to the one they have in vivo. They considered that the decline of the stimulation of prolyl-hydroxylase activity with increasing population doubling level and in fibroblasts of any doubling level incorporated in collagen lattices supported this view [Asselineau and Bell, 1984].

The problem whether the loss of the growth potential is an aging process has been discussed above (chapter I, II), and we feel that there is enough data supporting Hayflick's concept. On the other hand, whether it is in addition a differentiation process, deserves consideration and is certainly an interesting approach.

Remacle and co-workers [Raes and Remacle, 1983; Raes et al., 1983] have also approached the evolution of the fibroblast stem cell compartment with the differentiation concept. Based on morphology, they could distinguish two types of cells in hamster fibroblast cultures and concluded that about 2% of the cells differentiate at each doubling from phenotype I into phenotype II. When 70% of the life span is completed, there is both an increase in the differentiation rate into phenotype II and a rapid aging process of the type I cells which consists of a slowdown of their division rate. These investigators suggested that the arrest of growth results from a terminal differentiation mechanism for some fibroblasts, and, for others, from a stochastic aging process.

Besides the morphology, type II cells could be distinguished by their DNA content: 47% have a 4c DNA content, 44% a 8c and 6% a 16c DNA content. Only 3% at this stage are diploid [Raes et al., 1984].

An increase in ploidy in terminal human [Matsuo et al., 1982] and chicken [Kaji and Matsuo, 1981] fibroblasts has been reported (fig. 13, 14). It has been suggested that the polyploidy is due to endomitotic replication of DNA, unaccompanied by cell division, a process that would lead to terminal differentiation [Mulherkar and Sharma, 1979].

One of the important features of phenotype II hamster cells described by Remacle's group, is a different organization of the microtubules, with loss of polarity. This is expressed as a centrifugal depolymerization of the microtubules after treatment with nocodazole, contrary to type I cells where the depolymerization is centripetal. This phenomenon was later analyzed with more detail [Raes et al., 1984] and led to the conclusion that in the type II cells, microtubules depolymerize from the centrosome and not from the cell periphery as in other cells. The structure of the centrosome also appeared modified with a diminished pericentriolar material. The same authors proposed that in type II cells, there is an increase in the critical concentration of tubulin at the minus end of the microtubules, compared to the other cells. From these findings they concluded that what first appeared as differentiation is in fact the result of a degenerative process. In any case, the disturbance in microtubule organization must be an important element of the mechanisms triggering an increase in cell size.

Bayreuther and co-workers have extended the differentiation concept proposing [Konterman and Bayreuther, 1979; Mollenhauer and Bayreuther, 1986; Bayreuther, personal commun.] that there is a pluripotent stem cell which is the progenitor of the fibroblast, the myoblast and the chondroblast stem cell systems. Based on studies of morphology, DNA content, RNA synthesis, collagen synthesis, protein synthesis, protein content, two-dimensional gel electrophoresis of ^{35}S-methionine polypeptide pattern, and monoclonal antibodies, they followed the evolution of the putative unipotent fibroblast stem cell through a three stage differentiation sequence of the dividing skin fibroblasts of chicken, mouse, rat and man *in vitro* along the mitotic cell lineage, fibroblast cell type I–II–III. Mitotic fibroblast III would shift spontaneously to postmitotic fibroblast IV. In in vivo-in vitro experiments Mollenhauer and Bayreuther [1986] found in primary populations obtained from BN rat skin and lung a progressive shift to fibroblast types II and III as a function of the increasing animal donor age.

RNA synthesis was found to be higher in the postmitotic cells and collagen synthesis increased through the evolution of type I to type III fibroblast with a shift in the ratio of alpha1- and alpha2-chains of collagen type I [Konterman and Bayreuther, 1979; Mollenhauer and Bayreuther, 1986]. In bidimensional electrophoresis new spots could be seen in protein extracts from postmitotic cells.

In other words, according to Bayreuther and co-workers, the progression of the fibroblasts along the multistage differentiation sequence is directly correlated to the in vitro and in vivo age of the fibroblast population.

Macieira-Coelho and Lima [1972] followed in chicken embryo fibroblast population the changes in cell volume and found a three step evolution which would fit the progressive appearance of three different cell populations (fig. 37). The presence of three morphological cell types in human fibroblast populations, with an increase during serial divisions in vitro of types II and III, now seems well established. Indeed, several investigators have reported their presence [Franks and Cooper, 1972; Russell and Witt, 1976; Hennis et al., 1981; Steinhardt, 1985]. The relationship, however, of these cell types with aging and pathology is not yet clear. It has for instance been claimed that the 3 cell types of mitotic fibroblasts could be used to distinguish normal, keloid and hypertrophic scars [Conway, 1960]. A reapraisal of these findings [Russell and Witt, 1976], however, reached the conclusion that the relative proportion of the three types of cells changes during the growth of the cells in vitro, but it evolves with the same pattern in the three different tissues. On the other hand, Hennis et al. [1981] reported that major changes in the occurrence of the cells of different shapes were not apparent with increased donor age, albeit type I cells decreased slightly with donor age. The frequency of the different cell types in cultures from skin biopsies of human donors of different ages was also examined by Steinhardt [1985]. She found that phenotypes I and II fibroblasts were represented to a greater extent in skin fibroblast cultures from a 2-year-old child than from 30-, 82- and 92-year-old donors, but could find no difference between the adult donors. She suggested, though, that variability may increase with age. According to her results, the most striking feature was the scarcity of clones of phenotypes II and III in the fibroblast population from the 2-year-old donor and their poor growth as compared to the same clone types from adult donors. In other words, fibroblasts types II and III from adults seem to possess a higher division potential than those of infant or embryonic origin. Steinhardt related this finding with the claim of Bell et al. [1979] that fibroblasts are mainly involved in wound-healing in infants, but that in the adult organism fibrocytes also participate in that process. With the progressive decline of fibroblasts type I in the adult organism, the fibrocytes would become more receptive to the induction of proliferation.

Since it has been reported that UDS decreases during differentiation [Hahn et al., 1971], Hennis et al. [1981] and Vijg et al. [1986] measured

UDS in the three cell types obtained from skin biopsies of human donors of different ages and from rats. This was described in chapter III.2.b. Vijg et al. [1986] interpreted their results in terms of a differentiation process.

In regard to what happens in the nucleus, the appearance of extrachromosomal circular DNA in phase IV cells has also been observed in differentiating systems [Yamagishi et al., 1983]. Production of circular DNA, however, is also found in nondifferentiating cells in vitro and its significance for cell behavior is not yet understood [Rush and Misra, 1985].

Some metabolic events could support the concept that the evolution of the different cell types is a differentiation process. Thus, Takahashi and Zeydel [1982] found some analogies in enzymatic changes occurring in human embryonic lung fibroblasts in phase IV and in 3T3 cells differentiating to adipocytes. In both situations gamma-glutamyl transpeptidase, glutathionase and phosphate-independent glutaminase activities became severalfold higher. The ratio of reduced to oxidized glutathione also decreased in both cell systems. An increase in the level of the cAMP-dependent protein kinases in phase IV human fibroblasts and during terminal differentiation of 3T3 preadipocytes and of myoblasts, has also been reported [Liu et al., 1986]. On the other hand, Vogel and Pitcher [1982] could not find any evidence of specific extracellular matrix components, suggestive of differentiation.

Matsumura et al. [1979a] described during the period of growth decline of mouse fibroblasts, an enhanced production of the colony-stimulating factor for mouse bone marrow cells, a feature he interpreted as favoring a differentiation stage.

Another event in the nucleus that could be compatible with the differentiation concept is the change in DNA conformation. This is suggested from the decline in DNAase I sensitive sites [Dell'Orco et al., 1983], from the changes in the organization of the 30-nm chromatin fibers [Puvion-Duttileul et al., 1984] and from the different rates of digestion by micrococcus nuclease [Dell'Orco et al., 1986]. Changes in DNA conformation have been found to correlate with modulation of genetic activity [Davies et al., 1984; Reeves, 1984]. Although they cannot constitute a proof in favor of differentiation, they can at least explain many of the modifications in the synthesis of different macromolecules described above. The increased DNA topoisomerase I activity in the terminal cell is also compatible with variations in the conformation of DNA [Hamelin et al., 1984].

Kator et al. [1985], on the contrary, tested on the fibroblast system the hypothesis that aging is a process of dysdifferentiation [Cutler, 1982]. They attempted to detect a derepression of the expression of the alpha- and beta-globin genes during the serial divisions of human embryonic lung fibroblasts in culture. This approach was justified by the previous finding of Ono and Cutler [1978] of a twofold age-dependent increase of alpha- and beta-globin RNA in mouse brain and liver. Globin RNA, though, could not be detected above normal background noise levels in aged fibroblasts, so that the results did not support the hypothesis that these cells grown in vitro are undergoing a dysdifferentiation process [Kator et al., 1985].

In chick fibroblasts no changes in template activity of chromatin from cells at different states could be observed [Courtois, 1974], which was considered as a distinction between in vitro senescence and differentiation favoring the former.

The concept that the evolution through serial divisions of the fibroblast cell compartment is a succession of differentiating steps is appealing. It is obvious from the data described above that no definite proof in favor of differentiation has been provided so far. The hypothesis will remain speculative as long as a function in vivo, for the different cell types, is not found, as is the case for other cell compartments with a limited number of potential doublings.

Many of the functions of the fibroblast are known. The fibroblast is an ubiquitous cell in the organism and in this respect it is responsible for the creation of a microenvironment either through the synthesis of macromolecules (collagen, proteoglycans, elastin, structural glycoproteins) which act as supporting structures [Robert and Robert, 1973], or through the synthesis of small soluble diffusable molecules like growth factors [Clemmons and Shaw, 1983; Kurobe et al., 1985] and prostaglandins [Hunter et al., 1984] that are important for the regulation of the homeostasis of the tissue. Other substances secreted by fibroblasts are heparin-like molecules [Marcum et al., 1986], alpha$_2$-macroglobulin [Brissenden and Cox, 1982], albumin [Lemonnier et al., 1979] and thromboxane A_2 [Hopkins et al., 1978]. Thus, the fibroblast can be considered as a differentiated cell.

The need for the interactions between fibroblasts and other cell types during development is well known. Indeed, mouse embryo mammary rudiments fail to develop in the absence of mesenchyme [Kratochwil, 1969] and the response to sex hormones of the epithelia from the genital tract, is mediated via the connective tissue [Cunha, 1986]. Vice versa, epithelial cells have inductive effects on mesenchymal cell. The stimulatory effect on stromal DNA synthesis, of growing mammary ducts, is an example of this interaction [Berger and Daniel, 1983]; aged epithelium, however, fails to elicit this response.

Deregulations in the fibroblast metabolism on the other hand, can lead to disease. In pathologic situations, the interactions between the mesenchymal and the neighboring tissues can be one of the mechanisms triggering the disease. In the joints for instance it is presumed that an interaction between synovial cells and cartilage are of importance in the pathogenesis of arthritis [Peeter-Joris and Vaes, 1984]. Also in this regard, interferon-mediated prostaglandin E and hyaluronic acid synthesis by fibroblasts after viral infection could be implicated in the inflammatory process of the joint [Yaron et al., 1981]. In the lung, a monocyte inhibition of fibroblast growth, related to the stimulation of fibroblast prostaglandin production, may be involved in the development of pulmonary fibrosis [Elias et al., 1985]. On the other hand, thromboxane production by lung fibroblasts, may be a mediator of anaphylaxis and asthma [Feinmark and Bailey, 1982].

Furthermore, evidence has accumulated recently, suggesting that in neoplasia alterations of the mesenchyme may create a field effect favorable for the transformation of epithelial cells (see chapter II.2). Germane to this problem is the finding that conditioned medium obtained from cultures of human embryonic fibroblasts was necessary for the maintenance of differentiating myeloid leukemic cells [Collins et al., 1977].

Fibroblasts differ in their properties, depending on the organ or tissue they come from. An example of this differentiation is the finding that human skin fibroblasts synthesize one-tenth as much alpha$_2$-macroglobulin, a plasma protease inhibitor, as lung fibroblasts [Brissenden and Cox, 1982]. Indeed, in many instances, differentiation concerns quantitative rather than qualitative differences. These results, and others already mentioned and that will be further discussed in chapter VII, suggest that during development, the fibroblast evolves into different metabolic pathways and thus acquire a tissue specificity, adequate for the local homeostasis.

The fibroblast also plays an important role during wound healing [Gabbiani, 1979; Gabbiani and Rungger-Brändle, 1981]; it migrates and invades the wound on a fibrin layer, followed by the deposition of collagen which imobilizes the fibroblasts, so that the epithelial cells can move in to fill the wound. One important step in wound healing is the transformation of local fibroblasts into myofibroblasts which acquire properties typical of cultivated fibroblasts [Gabbiani, 1979] such as the presence of contractile proteins which are responsible for the contraction of granulation tissue. During this process, fibroblasts and epithelial cells become connected by gap junctions, suggestive of intercellular communications [Gabbiani, 1979].

Therefore, it is obvious that the fibroblast is at any time a differentiated cell, with many specific functions. But does it evolve in vitro and in vivo towards a terminal differentiation like for instance the keratinocyte?

The latter cell which also has a division potential inversely correlated with the donor's age, has been shown to proliferate in vitro towards the same end as in vivo. This has been elegantly studied in depth by Green and co-workers. Addition of growth factors can prolong the growth potential of human epidermal keratinocytes and delay terminal differentiation [Rheinwald and Green, 1977]. In addition to growth factors, these cells require other substances produced by a feeder layer of irradiated 3T3 fibroblasts. Regardless, however, of the size of the clonal expansion of the initial population plated in culture, the cells eventually lose their division potential and acquire the characteristic of terminal keratinocytes in vivo, i.e. they develop detergent-insoluble cytoplasmic filaments, there is formation of a cornified cell envelope and destruction of the cell nucleus [Green, 1977]. Cell-cell contacts are not required for the development of detergent insolubility and unlike proliferation, which is dependent upon fibroblast products, terminal differentiation proceeds in the absence of a feeder layer [Green, 1977]. In this cell system, at the end, there is cytoplasmic enlargement while the nuclear size and DNA content do not vary [Barradon and Green, 1985]. Beyond a certain size the cells cannot divide and cannot be rescued from terminal differentiation [Barradon and Green, 1985]. The increase in size is accompanied by the appearance of new proteins such as keratins [Fuchs and Green, 1980], involucrin [Watt and Green, 1981; Banks-Schlegel and Green, 1981; Rice and Green, 1979] and other specific proteins [Simon and Green, 1984; Resing et al., 1984].

Chondrocytes, which also have a division potential in vitro inversely correlated with the age of the donor, can behave in an opposite way when serially subcultivated, depending on the experimental conditions [Evans and Georgescu, 1983]. Monolayer cultures of chondrocytes appear to dedifferentiate after several divisions. However, when grown as clones with the methods developed by Green they retain better their differentiated properties. Their growth potential under both circumstances is the same. This cell culture system is not a terminal differentiating one, since the cells at the time of explantation are already differentiated. In other words, they are differentiated cells with a limited division potential. Differentiation was measured by the production of type II collagen, synthesis of type IV proteoglycan and cartilaginous material, and cellular morphology [Evans and Georgescu, 1983]. As these authors pointed out, loss of differentiated function may itself be a result of the ageing process. They cite in this regard the works of Weiss [1975] and of Silberberg et al. [1964] revealing the presence of atypical fibroblast-like cells in the articular cartilage of older animals. The aged cells seem to be atypical in function as well as in structure as suggested by the appearance of the pericellular matrix [Silberberg et al., 1964]. According to these investigators, the morphologically aberrant cells may represent the oldest chondrocytes in the joint.

So it is possible that in some cell compartments, aging could be a sort of deregulation of differentiation, i.e. a paradifferentiation. In this respect the role of the fibroblast could be that of creating the right microenvironment for differentiation to proceed in neighbouring cells. In other words, the inductive role of the fibroblast on neighboring tissues observed during embryonic development [Kratochwil, 1969; Cunha, 1986] would be maintained during the organism life span. In that case aging of the fibroblast may be one of the causes of this deregulation of differentiation of other cell types, during aging.

Embryonic chicken lens are another example of in vitro behavior of a differentiated cell compartment [Sanwal et al., 1986]. They undergo a terminal differentiation into fibers with structural chromatin modifications analogous to those in vivo.

Aortic endothelial cells, still represent another type of in vitro behavior in regard to differentiation in vitro. They have kinetics of proliferation during their in vitro life span, very similar to those of fibroblasts [Rosen et al., 1981]. These cells maintained as monolayers retain throughout their life span a specialized function of the tissue of origin, i.e. factor VIII antigen expression. They also retain the ability to take up intercellular silver nitrate stain, which is another caracteristic of endothelial cells [Rosen et al., 1981]. When they enter phase IV they enlarge but maintain cell contacts and a monolayering typical of endothelial cells.

In adrenocortical cells serially dividing in vitro, some changes in differentiated function occur early during the cells' life span without further major changes towards the end [Hornsby and Gill, 1981]. Hence, in this cell system as these authors stressed, fully differentiated from the start, there are different regulations that become apparent, but one cannot speak of dedifferentiation or further differentiation.

Another cell type, brain derived cells, that have been used for in vitro studies of aging, were first thought to be glial cells [Pontén et al., 1983] but were later shown to be of leptomeningeal origin [Rutka et al., 1986]. No data are available in regard to the way they evolve metabolically during serial proliferation.

So the behavior of these different cell types are probably the expression of the diversity taking place in the organism. They all have in common a finite growth potential that is inversely correlated with the age of the donor. But the role of these cell types on the remodeling of the human body taking place through the animal life span, is certainly different. In some, one can obviously study the terminal differentiation that takes place. In others, a deregulation of differentiation occur during serial divisions that must disturb the role of the respective cell compartment in the organism. Still in other cell systems, a loss of differentiation probably takes place through proliferation.

How this drift taking place in cell populations through division, eventually leading to a postmitotic state, influences aging is an open question; but it is clear that there are multiple ways. A different proliferative response has obvious implications for aging, but many other yet unknown functions must also play an important part. Tissue culture has been and will be an irreplacable tool to approach these problems.

VI. Cell Cycle-Related Events during Growth Decline

It is apparent from the results described above that multiple mechanisms must be implicated in the decline of the growth potential of these cells during serial divisions.

1. Contact Inhibition of Growth

One of the striking events is the decline in the maximal density at confluence when growth stops. This allows the construction of a survival curve (fig. 1) which is widely used to express the cell population life span and as a reference to correlate functional changes with the number of generations completed. The decrease in the maximal cell density was observed to be more pronounced in postnatal that in embryonic fibroblasts [Hayflick, 1965; Macieira-Coelho and Pontén, 1969] and in cells from old as compared with those from young donors [Schneider and Mitsui, 1976]. The study of the kinetics of the inhibition of proliferation, led to the conclusion that old cells may become more sensitive to the events that limit growth during cell crowding [Macieira-Coelho et al., 1966a]. The disappearance of filopodia in old cells has been suggested to be associated with this increased sensitivity [Stanulis-Praeger, 1986]. On the other hand, during growth stimulation of human fibroblasts by hydrocortisone, there is an increased cell surface blebbing [Stanulis-Praeger et al., 1986]. It coincides with the time when the hormone stimulates the production of a putative autocrine growth factor, responsible for the increase of the cells saturation density [Finlay et al., 1985].

Several parameters play a role in contact inhibition of growth, a fundamental property of normal cells which is progressively lost during cell transformation. We will discuss mainly those that have been studied in regard to cell aging.

One parameter implicated in this respect, is the $CaCl_2$ concentration. Praeger and Gilchrest [1986] found that an elevated $CaCl_2$ concentration increases the cell saturation density, that this effect is lost during cell senescence and that the loss is donor age dependent. An age-related modification in the calcium growth-dependence has also been reported with cultivated cells from the thigh muscles of rats of different ages [Swierenga et al., 1976].

The presence of growth inhibitors is also crucial for the growth arrest during cell crowding. Indirect evidence for the presence of a growth inhibitor in human fibroblast was obtained by Garcia-Giralt et al. [1970] who found that growth could be induced in confluent cells by repeated medium changes until the population entered a refractory stage where it was not inducible anymore. Supernatants at this stage were growth inhibitory. Growth could also resume if the cell layer was washed, which suggested that a diffusable factor was removed by stirring.

Moreover, Milo [1973] could isolate from old cells, a glycoprotein with 50,000 M_r which was inhibitory for early passage cells, and could induce premature aging. It seemed to be associated with the cell membranes. Other investigators also reported that plasma membrane glycoproteins from human fibroblasts are growth inhibitory and can induce a state similar to quiescence [Wieser and Oesch, 1986]. A high molecular weight glycoprotein found to accumulate at the end of the life span of normal human skin fibroblasts and in those from a patient with Spielmayer-Vogt syndrome [Poot et al., 1986] could belong to that group of molecules. Membrane glycoprotein changes are known to occur during senescence of normal human fibroblasts [Milo and Hart, 1976; Blondal et al., 1985]; either through increased production or through a different coordination in their synthesis, these molecules seem to play a role in the increased sensitivity to cell crowding.

Further evidence for the presence of growth inhibitors came from cell hybridization experiments which were described in chapter III.3. They suggested that a growth inhibitor whose production increases during senescence is an additional factor responsible for the increased sensitivity of old cells, to cell cycle inhibition during crowding.

Concerning the role of other proteins (statin) found associated with arrested cells, either reversibly arrested or postmitotic [Wang, 1985a, b], since no known function has been ascertained, it is for the moment premature to suggest a relationship of causality with the inhibition of growth.

Besides soluble diffusable factors, topological constraints are also limiting for cell division during cell crowding. They depend inter alia on the matrix where the cells attach and migrate. Indeed, it has been shown that contact inhibition of growth is substratum dependent [Macieira-Coelho and Avrameas, 1972].

Collagen is one of the molecules secreted by the cells, which is a component of the extracellular matrix, and whose role in the regulation of growth starts to be better understood. Collagen is inhibitory for cell movements and growth [Bard and Elsdale, 1986] and an increased production of this molecule has been reported in aged cells [Kontermann and Bayreuther, 1979]; a change in the ratio of the alpha-1/alpha-2-chains of the molecule also takes place.

Fibronectin is another important component of the extracellular matrix that is growth regulatory and it has been suggested that in vivo growth of fibroblasts depends on the equilibrium between collagen and fibronectin [Nagata et al., 1985]. The fibronectin of old cells is not as efficient in promoting adhesion as that of young cells [Millis et al., 1977]. This seems to be due to structural changes in the molecule [Sorrentino and Millis, 1984] since fibronectin from young cells increases the adhesion of old cells to the substratum and old fibronectin is less effective in supporting the adhesion of either cell type [Chandrasekhar et al., 1983a]. Fibronectin of aged cells, also binds poorly to collagen type I and II [Chandrasekhar et al., 1983b]. Although less fibronectin is found at the surface of the old cell, there is an increased production [Vogel et al., 1981; Shevitz et al., 1986] maybe because of the absence of a feedback control on its synthesis, since the molecule cannot anchor at the cell surface [Vogel et al., 1981]. Further evidence for changes in the regulation of fibronectin metabolism comes from the observation that the internalization process of the molecule is slower [Geuskens et al., 1986].

Proteoglycans are another class of molecules that are secreted by the cell, contribute to the composition of the extracellular matrix, and influence attachment and growth. Schachtschabel and Wever [1978], Wever et al. [1980], and Sluke et al. [1981] examined the synthesis and distribution pattern of glycosaminoglycans during aging of human embryonic fibroblasts. A reduced rate of incorporation of radioactive sulfate and glucosamine occurred only during the last 4–5 doublings, with a relatively greater decline of ^{35}S-sulfate incorporation than of ^{14}C-glucosamine into extracellular glycosaminoglycans. There is, however, a relative increase of ^{35}S-sulfate and ^{14}C-glucosamine incorporated into cell-bound and extracellular heparan sulfate. Treatment of young cells with heparin inhibited growth and also produced a relative increase of the radioactivity incorporated into heparan sulfate. In young cultures, heparin caused an overall increase in the synthesis rates of glycosaminoglycans. The comparison of the distribution of glycosaminoglycans in old versus young cells confirmed a reduced synthesis of total glycosaminoglycans in the former with a relative increase in heparan sulfate and a relative decrease of dermatan sulfate and hyaluronic acid synthesis. The synthesis of chondroitin sulfate did not vary.

The relationship between growth decline and surface proteoglycans was also investigated by Matuoka and Mitsui [1981]. They confirmed that the relative amount of heparan sulfate increases in old human fibroblast populations and that this change was inversely correlated with the saturation density. On the other hand, confluent glutaraldehyde-fixed, late passage cells were inhibitory for the growth of young cells; treatments that degraded heparan sulfate removed the inhibitory activity. The authors suggested that heparan sulfate is involved in the increased sensitivity to the various mechanisms inhibiting division during cell crowding.

Vogel et al. [1981b], however, came to the conclusion that glycosaminoglycan synthesis and composition are not related with this parameter; they found a general decline in the synthesis of sulfated glycosaminoglycans that are released into the medium but not of those located at the cell surface. Vogel et al. compared human embryonic fibroblasts aged in vitro and fibroblasts from young and old donors. More hyaluronic acid was produced and released into the medium by cells from the old than by those from the young donors; in addition, in old fibroblasts, there was a delayed movement of ^{35}S-glycosaminoglycans (GAG) from the cytoplasm to the surface and extracellular compartment. On the other hand, the amount of ^{35}S-GAG per milligram cell protein found at the surface and the composition of ^{35}S-GAG in each cell compartment did not vary. They concluded that secreted and cell-surface GAGs represent different pools and that in vitro aging has its effect primarily on the secreted pool.

Gordon et al. [1985] reported that early passage endothelial cells grown to confluence on a matrix from late passage cells incorporated less GAG into the extracellular matrix than when grown on a matrix of early passage cells. Also incorporation of GAG into the extracellular matrix was higher when late passage cells were grown on a matrix from early passage cells. Early passage cells contained more GAG and less heparan sulfate.

Although the situation is not completely clear yet in regard to the changes taking place in proteoglycans and to their role in cell growth, a disturbance in their synthesis and distribution certainly takes place and contributes to the alterations in cell-substratum adhesion and proliferation. In particular, heparan sulfate seems to be indeed a growth inhibitor that increases with cell senescence and a limiting factor for proliferation during cell crowding. Moreover, sulfated glycosaminoglycans play a role in the binding of fibronectin to collagen [Hedman et al., 1982] and thus must be also responsible, through this relationship, for the alterations in the extracellular matrix. One can infer that the modifications in cell-substratum adhesion are likely caused by a disturbance in the interaction of different molecules whose synthesis is differently coordinated and whose structure, at least in the case of fibronectin, may be altered.

The role for cell proliferation of the surface where the cells attach is also obvious from experiments showing that the final cell densities and the number of generations can be increased, growing smooth muscle cells on dishes coated with extracellular matrix [Gospodarowicz et al., 1981].

Internal topological constraints are also crucial for the regulation of growth. In this respect, an event that contributes to the decline in the maximal cell densities with aging is the increase in cell volume [Simons, 1967] (fig. 52). This was first detected by measuring cell diameters under a microscope with the aid of an ocular micrometer and was later confirmed with electronic counters [Macieira-Coelho and Pontén, 1969; Simons and van den Boerk, 1970; Mitsui and Schneider, 1976a; Whatley and Hill, 1979] and with measurements of the cell area on microphotographic paper [Bowman et al., 1975a; Greenberg et al., 1977; BeMiller and Miller, 1979].

Not only does the cell area increase, nuclear and nucleolar areas also increase, and the pattern of increase is similar for the three measures [BeMiller and Miller, 1979]. There is also a good correlation between changes in area and dry mass [BeMiller and Miller, 1979]. In cells of postnatal origin, the beginning of the increase in nuclear area was faster in cells from older donors [Lee et al., 1978].

The question of the relationship between volume change and the decline of the growth potential was studied by Macieira-Coelho [1970], who cocultivated young and old cells in different proportions and followed the growth and terminal cell densities of the mixed populations (fig. 57). The presence of the two cell types (young and old) was ascertained by the volume distributions of the separated and mixed populations (fig. 58). The results showed that the final cell density was inversely related to the number of large, old cells in the culture. It was concluded

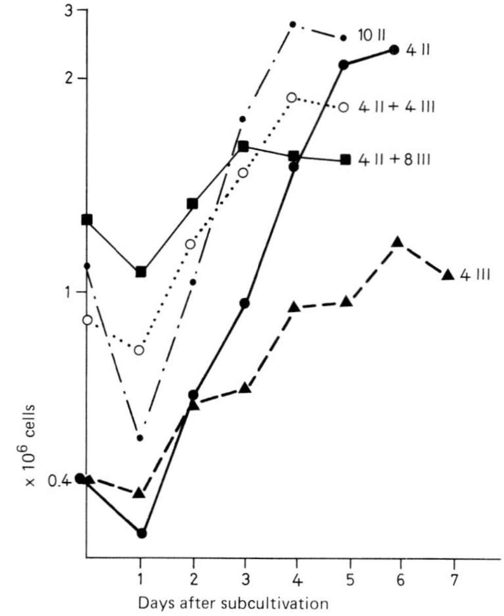

Fig. 57. Growth curves of phases II and III human embryonic lung fibroblasts grown separately and cocultivated after different inocula. One group received 10^6 phase II cells, another group 4×10^5 phase II cells, a third group 4×10^5 phase III cells, a fourth group 4×10^5 phase II and 4×10^5 phase III cells simultaneously and finally a 5th group 4×10^5 phase II and 8×10^5 phase III cells [Macieira-Coelho, 1970].

that the growth arrest at progressively lower densities in older cultures can be due to an increased sensitivity to contact inhibition of division and to a decreased surface area available for attachment, old cells being larger, need a wider area. Indeed, Blomqvist et al. [1978] found with another cell system (brain-derived cells) that the old cell is thinly spread over the substratum, with many focal points of contact and large areas of close opposition between cell and substratum.

So these works show that external and internal cellular constraints limit the final density that the cell population is able to reach. The response of the cell to external factors is modified, and the synthesis by the cell, of substances that create its own external microenvironment and that can have a negative or positive feedback effect on growth, is altered.

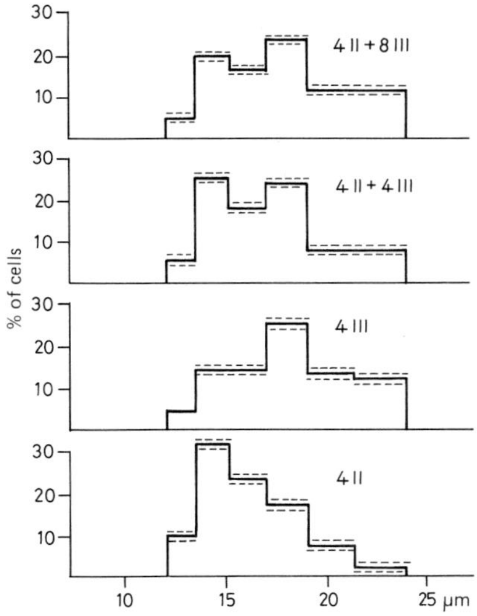

Fig. 58. Size distributions found on the first day after subcultivation in the same experiment as the one represented in figure 57. The dashed lines represent the 95% confidence limits [Macieira-Coelho, 1970].

Furthermore, there seems to be an excess production or decreased elimination of soluble factors that are growth limiting. Finally, the spatial changes created by the increased cell size also increase the sensitivity to cell crowding, both by altering the intercellular relationships and the interaction within the cell between its different compartments at the molecular and supramolecular level.

2. Commitment to Divide

Cells in resting phase that are triggered to divide, go first through a period called G_1 that lasts approximately 6 h, during which several metabolic events take place that lead to DNA synthesis (fig. 59). The time for the synthesis of DNA lasts about 7 h and is called S period. DNA synthesis is followed by another period called G_2, which lasts approximately 4 h, during which the cells prepare for mitosis which lasts 30 min to 1 h and ends the division cycle with the separation of the two daughter cells.

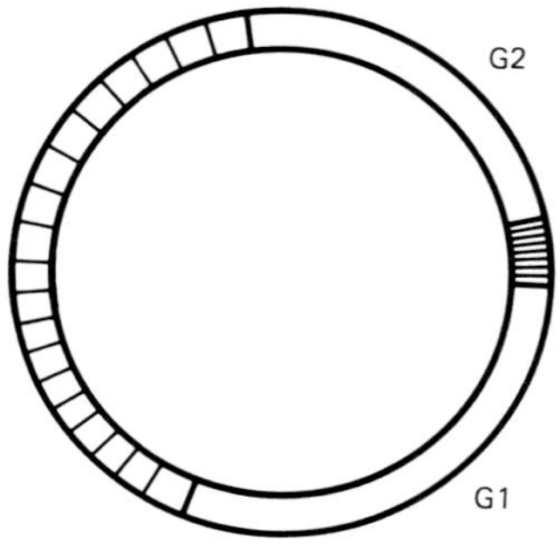

Fig. 59. Schematic representation of the cell division cycle. M = Mitosis; G_1 = interval between the end of mitosis and the initiation of DNA synthesis; S = DNA synthesis; G_2 = interval between the end of DNA synthesis and the initiation of prophase. The length of each segment corresponds to the approximate relative length of each period of the division cycle.

a. Initiation of the Division Cycle

Response to Growth Factors. One of the critical events necessary for the initiation of the division cycle is the competence to respond to growth factors. With aging in vivo cells become less responsive in the sense that higher concentrations of growth factor or more growth factors are needed to obtain a response. Adult human fibroblasts for instance, require ten times more platelet-derived growth factor (PDGF) than fetal fibroblasts [Slayback et al., 1977]. Adult human fibroblasts also have a lower proliferative response, than newborn fibroblasts, to thrombin, insulin, epidermal growth factor (EGF) and serum mitogens [Plisko and Gilchrest, 1983]. Cells from thigh muscles of fetal rats proliferate in vitro in medium supplemented with adult rat plasma [Swierenga et al., 1978]; as the animal matures, the cells become unresponsive to plasma but retain the ability to grow in medium supplemented with fetal bovine serum.

Furthermore, fibroblasts from patients with Werner's syndrome, which grow poorly in vitro, although they have a normal mitogenic response to fetal bovine serum, are defective in regard to adult serum, PDGF and FGF [Bauer et al., 1986]. Some PDGF-mediated pathways seem disturbed since this growth factor cannot induce collagenase synthesis as in normal fibroblasts. PDGF binding and the number of receptors, however, are normal in Werner's syndrome fibroblasts.

During in vitro aging, embryonic human fibroblasts require a higher concentration of dialyzed serum to obtain a given proliferative response [Ohno, 1979]. In regard to other growth factors (EGF, insulin, transferrin and dexamethasone), although the cells become less responsive, the concentration that elicits the maximum proliferative response does not change [Phillips et al., 1984]. With PDGF, however, they require increasing amounts to elicit the maximum mitogenic response.

The loss of responsiveness during in vitro aging of human fibroblasts, is selective: fetal bovine serum, tumor-derived factor, tumor promotors, EGF, dexamethasone, and microtubule-disrupting agents, stimulate DNA synthesis in quiescent cells, up to the time when 66–79% of life span is completed; when 94–97% of the life span is completed, only fetal serum, tumor-derived factor and EGF, can stimulate DNA synthesis; in the terminal stage (100% of life span completed) only SV40 can induce DNA synthesis [Tsuji et al., 1984].

The dependency on growth factors varies with the different cell types. The number of generations accomplished in vitro by endothelial cells seems to be very dependent upon the presence and concentration of growth factors [Johnson and Longenecker, 1982] and the different replicative life spans reported for these type of cells is probably due to differences in batches of serum. The division potential of fibroblasts seems to be less dependent upon these factors, since different laboratories have reported in general the same life spans. Indeed, EGF for instance, fails to alter substantially the division potential of human fibroblasts [Didinsky and Rheinwald, 1981], while it triples the number of generations of human keratinocytes [Rheinwald and Green, 1977] and increases several fold the generations of granulosa cells from bovine ovaries [Gospodarowicz and Bialecki, 1978]. The life span of bovine adrenocortical cells in culture is also prolonged by the presence of FGF [Hornsby et al., 1984; Hornsby and Harris, 1987].

Possible Mechanisms Involved in the Failure to Respond to Growth Factors. (1) Binding to receptors: The changes in the response to growth factors with age does not seem to be due to deffective binding. Rosenbloom et al. [1976] measured insulin binding to cultured human fibroblasts from normal donors and from patients with precocious aging and found increased binding with physiological as well as pathological aging. Increased binding of EGF to aged smooth muscle cells has also been reported [Bhargava et al., 1979]. This could be due to an increased number of receptors secondary to the changes in cell volume. On the other hand, Hollenberg and Schneider [1979] could not detect any change of insulin and EGF binding to fibroblasts from normal donors of different age groups. EGF binding ability is also stable during in vitro aging of human embryonic fibroblasts [Phillips et al., 1983].

(2) Supply of energy: The initiation of division involves several events. The first signals seem to consist in phosphorylation for the supply of energy and in changes of molecular conformation (topological constraints) visible at the cellular level by the induction of membrane movements. These two are certainly interdependent since the changes in conformation consume energy and may in turn increase the free energy available. Actually, the very first signal induced by growth factors could be a change in the conformation of the receptor, i.e. its allosteric transition, so that it acquires the kinase activity.

The basic role of growth factors is probably to make available the free energy necessary to start the chain of events, through phosphorylation of their receptors. Growth factor receptor kinases seem to be able to activate directly different metabolic pathways since several enzymes can serve as their substrates [Reiss et al., 1986]. It is possible that their efficiency in this respect varies to different degrees and that more free energy is needed during senescence due to the conformational modifications at the molecular and supramolecular level. It could also be that the chain of events along the cell cycle triggered by growth factors are differently affected during senescence and that progressively more steps along the chain are hampered. At the very end, as suggested by Tsuji et al. [1984], maybe only the viral gene products have the biological activities necessary to compensate for or bypass the different mechanisms defective in the terminal cell. An additional reason for the need of a stronger stimulus is the accumulation of an inhibitor whose action will become more difficult to overcome.

Qualitative modifications of the phosphorylation of proteins have been reported during cell senescence in vitro (though not in regard to growth factors) with reinforcement of the phosphorylation of some proteins and a relative decrease of others [Kahn et al., 1982]. These authors remarked that these modifications are progressive, as is senescence itself, and observed the appearance of new isozymic forms of protein kinases and new phosphoproteins. They speculated that these isozymes could correspond to new genes progressively activated during cell aging. So it is possible that this leads to a greater difficulty in making available the free energy necessary to initiate the division cycle, through a deregulation of phosphorylation processes. Another feature germane with this problem of energy mobilization is the finding that during in vitro aging, there is a decrease in ATP content after exposure to metabolic poisons [Muggleton-Harris and Defuria, 1985]. This was considered as an estimate of ATP turnover. The same investigators also found a decreased ATP turnover in cells from older subjects as compared to those of younger ones. These results are also suggestive of a deregulation in the mechanisms of phosphorylation.

Progressive changes in the growth factor-induced phosphorylation have never been reported. On the other hand, at the very end some disturbance may take place, since the autophosphorylation of the EGF receptor does not occur when membrane extracts are immuneprecipitated with antiserum-EGF receptor [Carlin et al., 1983a]. Autophosphorylation, however, occurs when the membrane extract is first phosphorylated and then immunoprecipitated [Carlin et al., 1983b]. This could mean that some deregulation takes place in the postmitotic cell with an associated factor necessary for the autophosphorylation, or that the receptor is modified so that the methodology used induces conformational changes not compatible with the autophosphorylation. This putative defect is a terminal event and thus, cannot explain the deregulation of the initiation of the cell cycle which occurs long before that last step. The PDGF-induced autophosphorylation, though, was reported to be unchanged up to the very end of the cells growth potential [Paulsson et al., 1986].

Another early event in the initiation of the division cycle is the activation of the proton pump across the cytoplasmic membrane; changes in topology at the molecular level would be enough to trigger the proton pump. To the best of our knowledge, cell aging has never been studied in this respect. Other mediators of energy transduction induced by growth factors have not been studied either.

(3) Topological constraints: One of the early effects of growth factors is the initiation of membrane movements [Baker and Humphreys, 1971; Brunk et al., 1976] which are followed by changes in the mass and volume of the nucleus [Mallucci et al., 1972]. This induction of movements declines during senescence; indeed one of the characteristics of the latter, as mentioned in the introduction, is the loss of cell motility.

A relationship between cell movements and the initiation of the division cycle has been acknowledged since the early years of tissue culture techniques. On the other hand, it was shown that growth can be modulated by membrane-substratum interactions [Macieira-Coelho, 1967b, c;

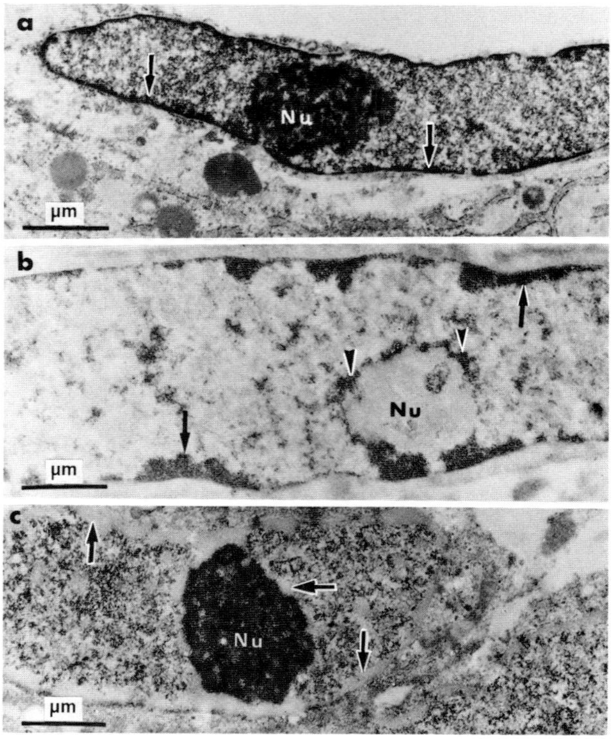

Fig. 60. Nucleoprotein organization following conventional fixation with glutaraldehyde alone of phase III human embryonic lung fibroblasts. ×18,000. Bar = 1 μm. *a* Uranyl and lead staining. Fibrillar and granular ribonucleoprotein components of the nucleolus (Nu) are intermingled. Chromatin is condensed at the nuclear periphery (arrows). *b* Specific DNA staining. Dense chromatin is accumulated at the nuclear (arrows) and nucleolar (arrowheads) peripheries. In addition, the reaction reveals dispersed chromatin within the nucleoplasm and the nucleolar body (Nu). *c* Preferential ribonucleoprotein staining. The bleached chromatin is mainly located at the nuclear periphery and around the nucleolus (arrows). The latter (Nu) and the nucleoplasm which contain ribonucleoprotein structures remain heavily contrasted [Puvion-Dutilleul et al., 1982].

Macieira-Coelho and Avrameas, 1972; Macieira-Coelho et al., 1974, 1978; Wahrmann et al., 1981; Sénéchal et al., 1984] and that the first signal, triggered at the membrane level in these situations of growth modulation by the extracellular matrix, is of a physical nature [Macieira-Coelho et al., 1974]. Furthermore, it was later shown that the initiation of the cell cycle seems to be dependent upon changes in cell shape [Folkman and Moscona, 1978]. The finding that one of the early effects of growth factors, is to induce metabolic changes in focal adhesion sites [Hiu et al., 1985] also supports the hypothesis that motion is necessary for the initiation of the division cycle.

How can this information, flow from the membrane, influence the genetic expression and trigger the initiation of the division cycle? As suggested by Puck [1977], the cytoskeleton could be the link through which flows this growth-regulatory information from the membrane to the nucleus.

Several experiments suggest that the initiation of DNA synthesis is coupled with microtubule polymerization and depolymerization. Drugs, like colchicine, that cause microtubule disruption, stimulate DNA synthesis [McClain and Edelman, 1980]. Conversely, drugs that stabilize cytoplasmic microtubules inhibit the stimulation of DNA synthesis by growth factors in anchorage-dependent cells [Crossin and Carney, 1981]. During its stimulatory action on DNA synthesis, colchicine, also acts upon other elements of the cytoskeleton, such as the cytoplasmic actin complex [Van Gansen et al., 1984] and a model of feedback loops has been proposed [Van Gansen et al., 1984] between the number of mitotic cycles, the rate of DNA synthesis, the extension rate of the plasma membrane and the microtubular complex.

Not all drugs that change cell morphology via the cytoskeleton stimulate division; such is the case of cytochalasin B [Crossin and Carney, 1981]. Because this drug is known to trigger nuclear expulsion from the cytoplasm, it is possible that the disruption of the relationship between the nucleus and the rest of the cell inhibits the initiation of the cell cycle. Indeed, the integrity of the elements composing the cytoskeleton (microfilaments, intermediary filaments, and microtubules) must be necessary for the maintenance of nuclear movements, which in many cases seem to be strictly related with the accomplishment of cellular functions [Belmont et al., 1980].

On the other hand, it has been suggested [Berezney and Coffey, 1977] that changes in the organization of the nuclear matrix are coupled with nuclear functioning. This is not surprising in the presence of the increasing amount of evidence showing that the matrix is the site of anchorage of DNA; not only DNA synthesis initiation sites seem to be fixed on the matrix [Berezney and Buchholtz, 1981] but also the activation of genes could be dependent on attachment of DNA to the nuclear cage [Jackson and Cook, 1982].

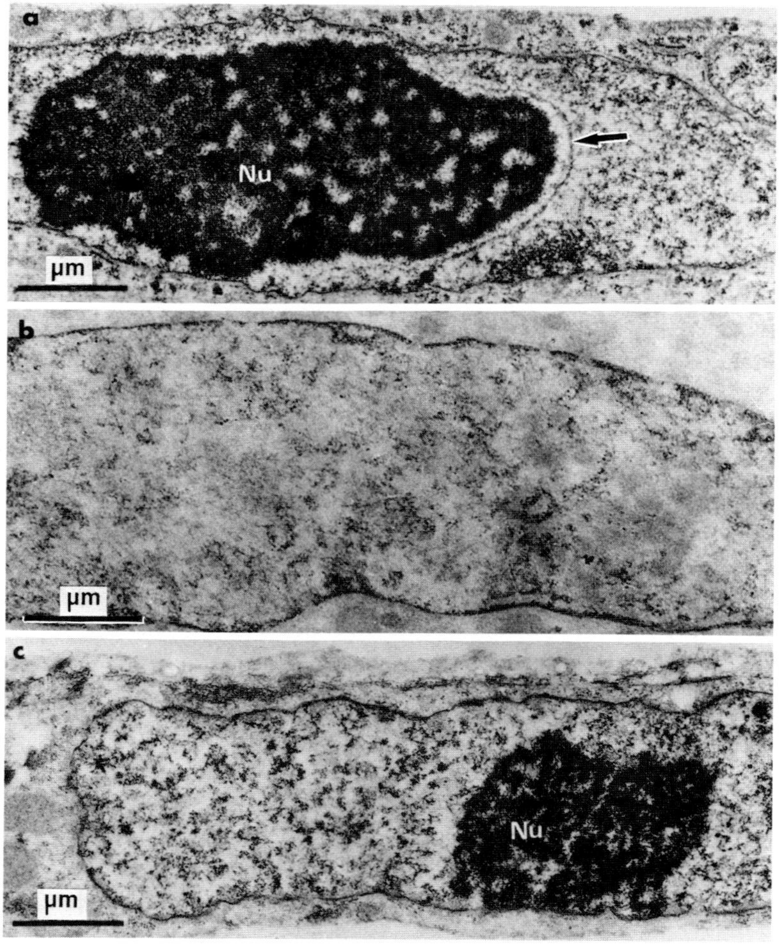

Fig. 61. Nucleoprotein organization following conventional fixation with glutaraldehyde alone of phase IV human embryonic lung fibroblasts. ×18,000. Bar = 1 μm. *a* Uranyl and lead staining chromatin is dispersed in the nucleoplasm which appears clearer than in early passages. The nucleolus (Nu) is prominent and presents a bud (arrow). *b* DNA-specific staining. Chromatin is largely dispersed in the nucleoplasm. *c* Preferential ribonucleoprotein staining. The nucleolus (Nu) is well contrasted, whereas non-nucleolar ribonucleoprotein structures are more dispersed than in early cultures [Puvion-Dutilleul et al., 1982].

Thus, it seems possible that nuclear movements are necessary for genome activity through modulation of chromatin conformation. Based on this assumption, it was suggested [Macieira-Coelho, 1983] that the first event that triggers the cell cycle is of a topological nature, so that membrane movements transmitted through the cytoskeleton to the nuclear cage help to give to DNA the right conformation for gene expression during the G_1 period, and to initiating sites the right steric position for the synthesis of DNA. Too much stretching of the cell or conversely, contraction of nuclear structures, may lead to the arrest of cell division. Excessive stretching is found in cellular senescence and contraction takes place during contact inhibition of division. In both instances, the probability of the chromatin structure to meet the right conformation and of DNA synthesis initiating sites the right steric position would be low and could decrease the probability of entering the cell cycle. Conversely, rapid movements will increase the probability for the initiating sites to reach the configuration favorable for DNA synthesis. This hypothesis is compatible with the concept of a chromatin cycle starting in G_1, which was first suggested by Mazia [1963]. Evidence in favor of this cycle has been obtained by different investigators [Pederson, 1972; Collins, 1977; Prentice et al., 1985].

In summary, motion and function are interdependent in a cell through the tension in a network of structures that extend from the periphery to the nuclear cage and that constitute the cellular scaffolding. The function of this network is regulated through the synthesis of molecules with the right steric configuration and through energy turnover. In the cell systems described herein, adhesion to a substratum is the trigger to build up tension. Cell behavior is determined by the way in which this network of structures is connected, i.e. by its topology, not by its particular shape and size. Motility is closely linked with cell division because it can create new topological constraints; in the absence of movements there may be a shift from division to other functions.

Concerning specifically the model of aging dealt herein, profound changes have been found at different levels of the cell, that can explain the deregulation taking place in this flow of intracellular communication.

As mentioned above there is a considerable amount of evidence showing that during cell senescence, there are modifications of cell adhesion. This must be related in part to the decline of the net negative surface charge [Bosman et al., 1976] which follows a pattern similar to that of the delay of the commencement of division after seeding the cells on a substratum (fig. 26, 45). The decline of the net negative surface charge is probably dependent on the decrease of membrane-bound sialic acid [Milo and Hart, 1976].

On the other hand, the matrix secreted by the cell is also fundamental for the creation of the right adhesion for the initiation of the division cycle. The composition of the extracellular matrix is also influenced by growth factors; indeed, they induce the synthesis of sulphated polysaccharides that are constituents of the extracellular space [Wasteson et al., 1973; Engström-Laurent et al., 1985; Westermark and Wasteson, 1975]. In particular, endothelial growth factor was found to decrease the amount of heparan sulfate [Gordon et al., 1985]. On the other hand, it stimulated the incorporation of sulfate and glycosamine into GAGs, hyaluronic acid, chondroitin sulfate and dermatan sulfate. Cell adhesion was enhanced. So the failure to respond to growth factors is also responsible for a defective matrix and poor adhesion.

Changes in the organization of cytoskeleton elements also take place. Cell spreading is altered and seems to be related with a disorganization of the assembly of actin bundles [Kelley et al., 1980] and a reduction of filamin which is implicated in actin-membrane associations [Kelley et al., 1985]. An increased actin content [Anderson, 1978] and a rearrangement of the microtubular network [Van Gansen et al., 1984; Raes et al., 1984] and of intermediate filaments [Wang, 1985b, c] have also been reported, which contribute to the decline in cell contractility [Azzarone and Macieira-Coelho, 1984] (fig. 48). A decreased cell spreading has also been observed in fibroblasts from normal aged and Alzheimer donors [Peterson et al., 1986].

The role of cell attachment and spreading on the initiation of DNA synthesis and its deregulation during cell senescence, is apparent when one compares the fraction of cells synthesizing DNA during the first 24 h after seeding the cells on a supporting surface, with the maximum number of cells capable of synthesizing DNA during a 24-hour period thereafter; while the former declines progressively, the latter remains fairly constant during most of the cells life span (fig. 26). These experiments show that when cells have built up the attachment sites and layed the matrix, they can enter division at the same rate, during most of their life span; with senescence, it takes longer to build up these sites and to attach and spread, so that the initiation of the cell cycle is delayed after inoculating the cells.

Further proof for the connection between the initiation of the cell cycle and the intracellular spatial modifications comes from studies showing that there is a relationship between the increase in cell volume and the progressive delay in the initiation of the division cycle. In chicken fibroblasts changes in cell volume during the cell population life span were found to precede disturbances in DNA synthesis [Macieira-Coelho and Lima, 1973] (fig. 37, 38). Furthermore, tritiated thymidine labeling of separated cells [Bowman and Daniel, 1975], direct analysis by cinematography [Absher and Absher, 1976] and comparison of cell volume with population doubling time [Mitsui and Schneider, 1976b], all led to the conclusion that larger cells have a decreased probability of entering the division cycle. It was further found that the increased cell volumes are directly related to the concomitant decline in cell replication rates [Mitsui and Schneider, 1976b]. Not all cells in an old population, however, are enlarged. A significant fraction have smaller sizes [Milo, 1973] and reversion from large to small size seems possible [Mitsui and Schneider, 1976b], probably through division of the large cell. The small cells could correspond to that fraction present in old populations known to have the same cycling characteristics as young cells [Kapp and Klevecz, 1976; Macieira-Coelho and Azzarone, 1982]. Also with respect to the relationship between size and probability of division, Collins et al. [1979] found that growth factor concentration and substratum areas necessary for proliferation increase with aging of brain-derived cells.

Changes in cell volume and probability of cycling were also compared with other types of behavior by Bowman and Daniel [1975]; these studies revealed that large cells are less motile, lack prominent bundles of microfilaments and have a decreased ability to make cell-cell contacts. Hence, it seems that decreased proliferation, increase in size, decreased attachment and decreased membrane movements are closely related. In spite of all these changes, the capacity of colchicine to stimulate DNA synthesis and to act upon the cytoplasmic microtubular and actin complexes is independent of the age of the cell [Van Gansen et al., 1984].

These spatial modifications have repercussions on the nuclear cage and must contribute to the different organization of the 30-nm chromatin fibers (fig. 24) and to the decondensation of chromatin seen under electron microscopy with conventional fixation (fig. 60, 61).

It is generally accepted that the 30-nm solenoid chromatin structure is organized in loops anchored to the nuclear lamina and projected to the center of the nucleus [Brasch, 1982]. The length of these loops can vary under different physiologic conditions [Brasch, 1982]. The labeling of cells with ^{3}H-thymidine and the observation of the distribution of the label on the 30-nm fibers favor this organization of chromatin (fig. 62). Indeed, two types of localization of silver grains occur, with an increased number of grains after longer labeling periods. Silver grains are in the main either scattered within the nonnucleolar area or are located at the nuclear periphery. This distribution of the silver grains are not modified whatever the duration of the labeling and the presence or absence of a long chase period. The peripheral labeling is not located at the level of the lamina densa, but rather at a short distance.

The progressive decrease in the density of the 30-nm chromatin fibers in the proximity of the lamina densa (fig. 24, 25) occurs in parallel with the delay in the initiation of DNA synthesis that follows attachment and spreading of the cell (fig. 26). These data support the view that topological modifications at the level of the high order structure of DNA, in particular at the level of the anchorage of DNA, are associated with the decline of the plasticity of the cell scaffold and the decrease in the flow of motion from the membrane to the nucleus (fig. 51).

(4) Cell-to-cell communication: Another event that controls the probability of initiating the division cycle and also depends on cell movements is cell-to-cell contacts. Growth factors influence this event indirectly, since they stimulate cell migration [Westermark and Blomquist, 1980]. Contact stimulation of division has been often neglected but is now well documented [Ohno and Yamada, 1984]. It probably is important to induce division through metabolic cooperation. Young, vigorous cells may this way cause the initiation of the cell cycle in older cells, where some metabolic pathways may take longer to be induced. Because a decreased motility will lead to a decline of the frequency of cell-cell collisions, it will also lower the probability of entering the cycle. But the decrease in cell contacts during aging is not only due to slower movements but also to a decline in intercellular adhesion, a change that must hinder cell-to-cell communication. A decline in intercellular adhesion has indeed been reported [Azencotte and Courtois, 1974; Spataro et al., 1979].

Additional evidence showing a defect in cell-cell contact phenomena comes from the work of Kelley et al. [1979] who found that gap junctions are sparser and smaller and that the time needed for the formation of gap junctions is prolonged in old cells. Correspondingly, there is a delay in the arrangement of particle arrays typical of forming junctions. The difficulty of old cells in establishing gap junctions probably explains the

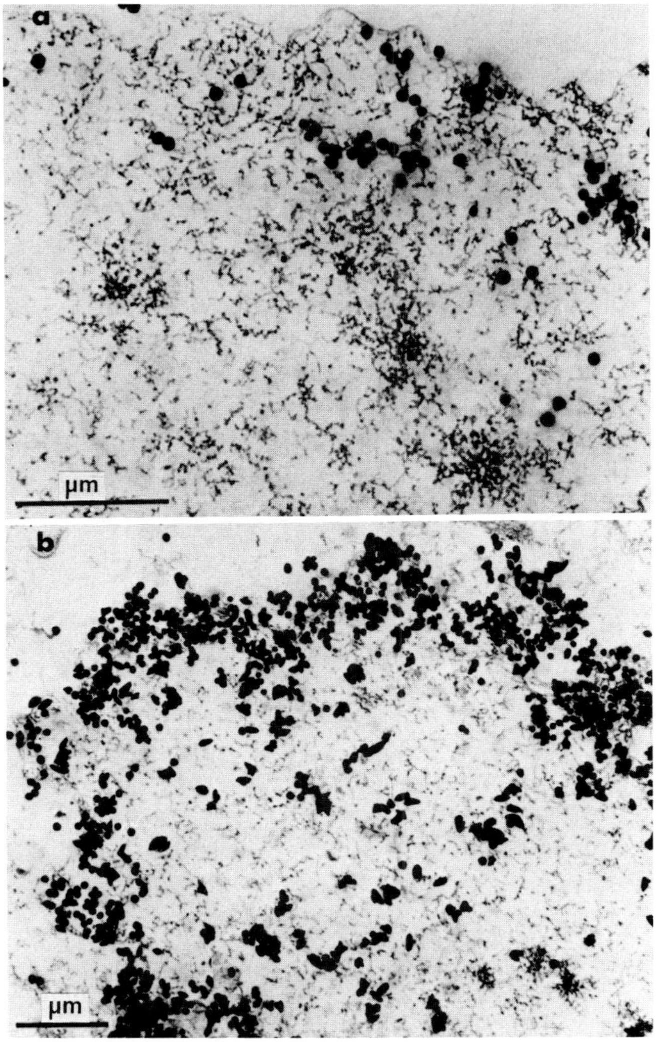

Fig. 62. Autoradiograms of phase IV fibroblasts incubated with ^{3}H-thymidine for 10 min *a* and 66 h *b* prior to the preparation of cells for the visualization of the 30-nm chromatin fibers [Puvion-Dutilleul et al., 1984].

reduced metabolic cooperation as measured by the exchange of radioactive precursors [Kelley et al., 1979; Kelley and Perdue, 1980].

(5) Protein metabolism: Another property of growth factors is to act as inhibitors of protein breakdown [Ballard and Read, 1985]. During in vitro senescence EGF and insulin maintain this property but lose the ability to stimulate protein synthesis [Ballard and Read, 1985]. These investigators suggested that the failure to induce DNA synthesis may be the consequence of the inability of senescent fibroblasts to achieve a rate of protein synthesis that exceeds protein breakdown.

Polyamines seem to play a role in the initiation of the cell cycle, although their mechanism of action is not yet understood. Changes in their metabolism have been reported during cellular senescence. A reduction of ornithine decarboxylase activity has also been found associated with in vitro senescence [Duffy and Kremzner, 1977; Chen et al., 1986]. It may be responsible for the reduced formation of putrescine or its metabolites such as spermidine after serum stimulation of quiescent embryonic cells aged in vitro or of progeria cells [Chen et al., 1986]. The induction, however, of ornithine decarboxylase by EGF reached comparable levels in young and old cells [Chua et al., 1986]. Polyamines can be incorporated into proteins via two metabolic pathways, one, leads to a 18,000-dalton protein and decreases in the old cell; the other pathway, which on the contrary is amplified, utilizes amino acids from putrescine via the polyamine catabolic pathway [Chen and Chang, 1986]. These switches in the pathways of protein synthesis, must also be implicated in the defective response to growth factors and the decreased probability of cycling during senescence.

DNA Precursors. Another mechanism that is important for the transit through the G1 period and the preparation for DNA synthesis and is disturbed during senescence of dividing cells is the availability of purine and pyrimidine bases through either de novo synthesis or reutilization of molecules (salvage pathway). In the presence of purine bases, the levels of phospho-ribosyl-pyrophosphate (P-Rib-PP) in normal cells are quite low because of the preferential utilization of the salvage pathways. In the absence of purine bases, the level of P-Rib-PP increases and the level of nucleotides decreases, both resulting in the stimulation of de novo purine synthesis. An age-dependent decreased response to deprivation of purine bases, both with respect to P-Rib-PP concentration and de novo purine synthesis has been described [Torrelio et al., 1982]. These findings are in agreement with a report from the same group [Paz et al., 1981] showing an age-dependent decrease in the ratio of two enzyme activities in the purine salvage pathway, the hypoxantine-guanine phosphoribosyltransferase, which is X-linked, and the adenine phosphoribosyltransferase, which is autosomal linked. The more extensive relative decline in the former enzyme was attributed to a gene dosage effect subsequent to the random inactivation of one X chromosome. The product of a single gene could become more vulnerable to the genome reorganization taking place in dividing cells. Hence, these changes could be the expression of a difficulty of the cells to adjust the P-Rib-PP concentration and the rate of nucleotide synthesis for the initiation of DNA synthesis.

Germane to these studies is the finding that glycinamide ribonucleotide synthetase levels, the second enzyme in the de novo purine synthesis pathway, increase during the growth decline of human postnatal human fibroblasts [Hards and Patterson, 1986]. It is possible that the decline in the salvage pathway is accompanied by a compensatory increase of the de novo synthesis. These investigators attempted to see if a gene dosage effect could be responsible for the increased enzyme levels. They compared diploid with monosomic and trisomic 21 cells and found that the increase took place to the same extent in all cell cultures although the population doubling level at which the enzyme activity started to increase varied for the individual cell lines. The results showed that the expected gene dosage effect was not observed.

The metabolic pathways that supplie the building blocks for the synthesis of DNA seem to be disturbed at other points during senescence. The activity of ribonucleotide reductase, the enzyme responsible for the reduction of ribonucleotides to deoxyribonucleotides, is decreased [Dick and Wright, 1982]. The pools of deoxyribonucleotides are also decreased [Dick and Wright, 1985b]. The decreased enzyme activity seems to be due to alterations of the structure of the enzyme, since inhibition of the enzyme by dATP follows the linear model with young cells, while the enzyme from old cells is inhibited according to a hyperbolic model [Dick and Wright, 1985a]. In the latter, the binding of the inhibitor to the enzyme does not stop completely the production of the product and the reaction therefore cannot be driven to zero.

So the events leading to the initiation of division and the transit through G_1 are quite complex and are just beginning to be understood. Due to these multiple mechanisms, there is a delay during cellular senescence in initiating the G_1 period and in the transit through G_1.

As mentioned in the description of the events taking place during phase IV, it has been claimed that terminal cells, when stimulated by growth factors, are able to proceed through G_1, but stop at the G_1/S boundary. But long before this last step, a difficulty in reaching DNA synthesis becomes apparent [Macieira-Coelho and Azzarone, 1982]. We believe that it is the understanding of these mechanisms creating the difficulty in reaching DNA synthesis, that is relevant to explain the deregulation of the control of cell growth that takes place in the organism.

b. DNA Synthesis

Once DNA synthesis is initiated it seems that it proceeds without any alterations during senescence [Griffiths, 1984; Hasegawa et al., 1985]. A report by Petes et al. [1974] claiming a retard in DNA chain elongation during aging of normal cells, measured with DNA fiber autoradiography, could not be confirmed [Hasegawa et al., 1982, 1985; Griffiths et al., 1983]. In cells from Werner's syndrome patients, however, a retarded rate of DNA elongation could be detected from the sedimentation profile of newly synthesizing DNA in alkaline sucrose gradients [Fujiwara et al., 1977], albeit not with DNA fiber-autoradiography [Hanaoka et al., 1983]. The latter technique could, however, detect an increase in the center-to-center distance between replicon initiating sites [Takeuchi et al., 1982; Hanaoka et al., 1983]. This was interpreted as an altered frequency of replication initiation in the cells from patients with this disease.

At the chromosomal level, it was found that the centromeric region of chromosome 9, terminates replication earlier at high population doubling levels of human embryonic fibroblasts [Lindgren and Farber, 1982]. This region is known to be rich in heterochromatin. It has also been reported that DNA synthesis appears to be reduced in the proliferating cells of aging cultures at the site of replication on the nuclear matrix [Collins and Chu, 1985]. This has been attributed to alterations in DNA structure such as an increase in the number of S1-nuclease-sensitive sites and breaks. They could also be due to the age-related changes in DNA conformation suggested from the increased resistance to digestion by DNAse I [Dell'Orco and Whittle, 1982] and from the rate of digestion by micrococal nuclease [Dell'Orco et al., 1986]. Other changes at the ultrastructural level, suggestive of conformational modifications that could modify the initiation of DNA synthesis in the old cell [Puvion-Dutilleul et al., 1984], are described in the preceding section. Since the steric position of initiating sites and the general conformation of DNA must be important for its synthesis, it is not surprising that at the very end, when the structural changes are profound [Puvion-Dutilleul et al., 1984], the probability of initiating DNA synthesis is very low. Still, the cells with profound chromatin structural changes can synthesize DNA, as seen by ultrastructural autoradiography after ^{3}H-TdR labeling; the distribution of labeling in these altered cells is identical to that of cells with young type chromatin [Puvion-Dutilleul et al., 1984]. This distribution is compatible with complexes of replication attached to the peripheral nuclear matrix.

The striking feature is the final abrupt, catastrophe-like, decline in the probability of initiating DNA synthesis [Macieira-Coelho and Azzarone, 1982]. Besides the dramatic changes in the organization of chromatin fibers which accompany this final event other phenomena occurring in the nucleus could explain this decline. One is the appearance of DNA in an extrachromosomal form [Kunisada et al., 1985] which occurs only in cells that reached the terminal stage [Icard-Lipkalns et al., 1986]. It is possible that transposable elements detach from and reintegrate into chromosomes at each cell division; at the very end, these elements may not reintegrate the chromosome because of the changes in chromatin conformation, so that the probability of synthesizing DNA is dramatically disturbed.

But other phenomena could also have a dramatic effect on the initiation of DNA synthesis, like the pronounced increase in 5′-nucleotidase [Sun et al., 1975; Remacle et al., 1980], an enzyme which hydrolyses nucleoside triphosphates and could this way interfere with the availability of the DNA building blocks. During the last divisions which we called phase IV [Macieira-Coelho and Taboury, 1981], as described in another chapter, many complex events take place; so the rapid decline in the probability of cycling must be due also to other parameters.

c. The G_2 Period

The first works that analyzed the changes in the cell cycle during in vitro aging found that when growth inhibition takes place during crowding in young cultures, cells start to accumulate in the G_2 period [Macieira-Coelho et al., 1966a, b]. This becomes more pronounced in older cultures, which have a fraction of cells with long G_2 periods before they reach contact inhibition of growth [Macieira-Coelho, 1967a, 1973b]. This accumulation of cells in G_2 was first interpreted as a block during this period of the cell cycle; subsequently, comparison between the data on DNA synthesis collected by autoradiography and that on DNA content obtained with cytophotometry, revealed that most of the cells are not blocked, and after variable times in G_2 reach mitosis. These studies led to the development of methods capable of revealing transit times [Macieira-Coelho, 1973b] and not only instants as before. They buttressed the view discussed above that what is relevant for the study of cell proliferation during in vivo aging is not the evaluation of the number of arrested cells, but rather the evaluation of how cells move along the cycle. It is the latter that expresses what happens during aging in vivo, in regard to the regulation of growth. These results concerning the accumulation of cells in the G_2 period were later confirmed [Kapp and Klevecz, 1976].

Subsequently, it was found that postnatal cells in vitro have a prolonged G_2 period, as compared to embryonic cells, suggesting that these changes are also related with developmental processes [Macieira-Coelho and Pontén, 1969].

Yanishefsky et al. [1975] used the technique of premature chromosome condensation to evaluate the period of the division cycle where cells are arrested. They found that only cells labeled with ^{3}H-thymidine induced G_2-type chromosome condensation and that 96% of the non-synthesizing DNA cells induced G_1-type chromosome condensation. Hence, these experiments suggest that most of the arrested cells are in the G_1 period.

The problem of the time spent by cells in vivo in the G_2 period has been extensively studied by Gelfant and Smith [1972]. They found that in the tissues, there are cells that start cycling from the G_2 period and that this fraction of cells is increased in old animals. Although we do not agree for the reasons described in chapter IV, with the concept proposed by Gelfant of noncycling cells for the explanation of aging of dividing cell compartments [Gelfant and Smith, 1972], his work and ours show an analogy between aging in vivo and in vitro in regard to disturbances occurring in the G_2 period.

Transit times in the G_2 period have been analysed not only in mesenchymal cells. Kubbies et al. [1985] found that lymphocytes from a 75-year-old donor stimulated by phytohemaglutinin (PHA) are arrested in the first S and G_2 phases, but not those of younger donors. Staiano-Coico et al. [1982], maybe because they did not use such sophisticated techniques as Kubbies et al., [1985], could not see any cells from old donors delayed in G_2, albeit the number of stimulated lymphocytes was lower. Lymphocytes from old donors, though, stimulated by PHA showed an increased sensitivity to the radiobiological effects of ^{3}H-thymidine, which is expressed by a higher percentage of cells arrested in the G_2 or M phase. So it seems that with aging, changes take place in the G_2 period which render the cells more vulnerable to radioactivity.

It is pertinent to remind in this respect works showing that during aging of human fibroblasts, in a significant fraction of cells, DNA is not distributed evenly between daughter cells [Macieira-Coelho et al., 1982], a phenomenon that occurs in the G_2-M phase. Toxic levels of ^{3}H-thymidine increased the fraction of cells where DNA is not evenly distributed.

All the events that take place during the G_2 period are not yet elucidated. One, however, that has been identified, seems to be implicated in the aging process. DNA synthesis proceeds by stretches (replicons) which are then bound end-to-end. The maturation of this gap-filling step between adjacent replicons takes place during G_2 [Van't Hof, 1980]. The following experiments suggest that one of the events responsible for the fragility of chromatin during cellular senescence, is a defect of the gap-filling step during the rejoining of adjacent replicons [Icard-Liepkalns and Macieira-Coelho, 1982].

When confluent human fibroblasts in resting phase are stimulated by the addition of fresh nutrient medium, a fraction of cells go more or less synchronously through one division cycle. Cells stimulated this way were labeled 8 h later, i.e. at the beginning of the S period, with ^{14}C-thymidine, and lysed at different times afterwards, at 39 °C, on top of a linear 5–20% alkaline sucrose gradient. Then they were centrifuged for 2 h at 40,000 rpm in an SW41 Beckman rotor. Figure 63 illustrates the sedimentation velocity of DNA at different times after adding the radioactive precursor to young and old cells. In both cultures, 45 min after labeling, the radioactivity is distributed in small peaks through the gradient which correspond to DNA intermediate in size between Okasaki fragments and bulk DNA. Then as these fragments became ligated, DNA progressively sedimented in young cells as a larger peak which is completed 10 h after labeling and corresponds to bulk DNA. In old cells, however, even at the 16th h after addition of ^{3}H-thymidine, DNA sediments in several peaks with different sedimentation velocities.

To see if the small molecular weight DNA in old cultures could be due to a larger fraction of cells that had not completed DNA synthesis because of a slow transit through the division cycle, identical cultures stimulated as those described in the preceding experiment, were fixed for autoradiography 10 and 16 h after labeling for young and old cells,

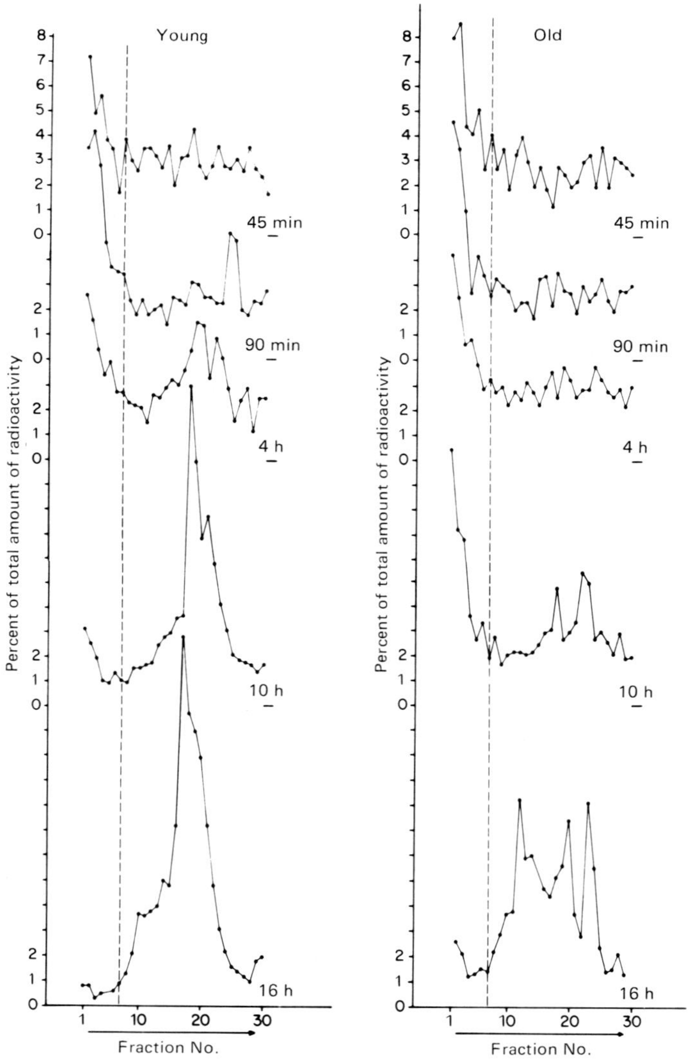

Fig. 63. Radioactivity collected from the different fractions of alkaline sucrose gradients, expressed as percentage of the total amount of the radioactivity of the sample. The vertical dashed lines indicate the fraction where denaturated lambda phage DNA (40.5 S) sedimented. The sedimentation direction is from left to right as indicated by the arrow. Young cells were at the 27th population doubling and old cells were in phase IV [Icard-Liepkalns and Macieira-Coelho, 1982].

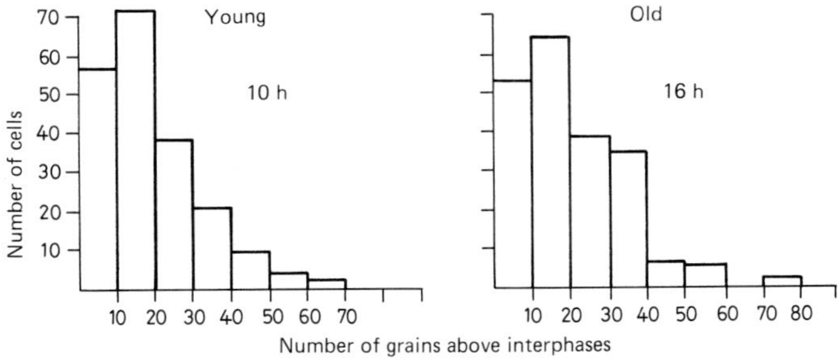

Fig. 64. Distribution of the number of grains above interphases in cultures identical to those used in the experiment illustrated in figure 63, and fixed 10 and 16 h after labeling of young and old cultures, respectively [Icard-Liepkalns and Macieira-Coelho, 1982].

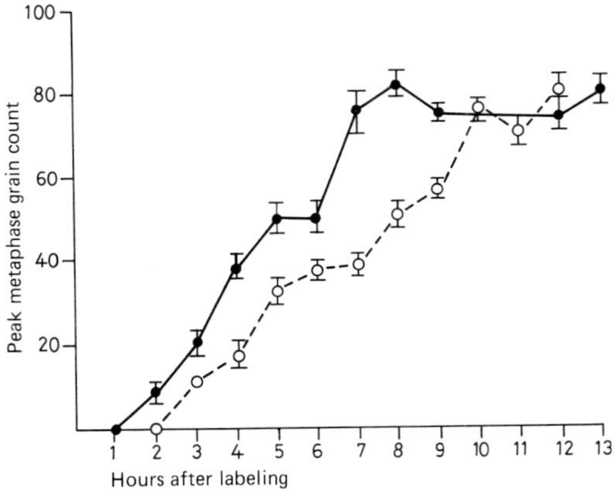

Fig. 65. Peak metaphase grain count determined in young and old cultures [Icard-Liepkalns and Macieira-Coelho, 1982].

respectively. Grain counts above interphases showed that the distribution of radioactivity was the same in both cell populations (fig. 64). The length of the S period was also measured in young and old cultures (fig. 65), evaluating the peak metaphase grain count at different times after adding ^{3}H-thymidine [Stanners and Till, 1960]. DNA synthesis was found to have been completed 7 and 10 h after labeling in young and old cells, respectively. Hence, the results show that in old cells in the G_2 period, DNA is thermo- and alkali-labile and sediments in a dispersed fashion.

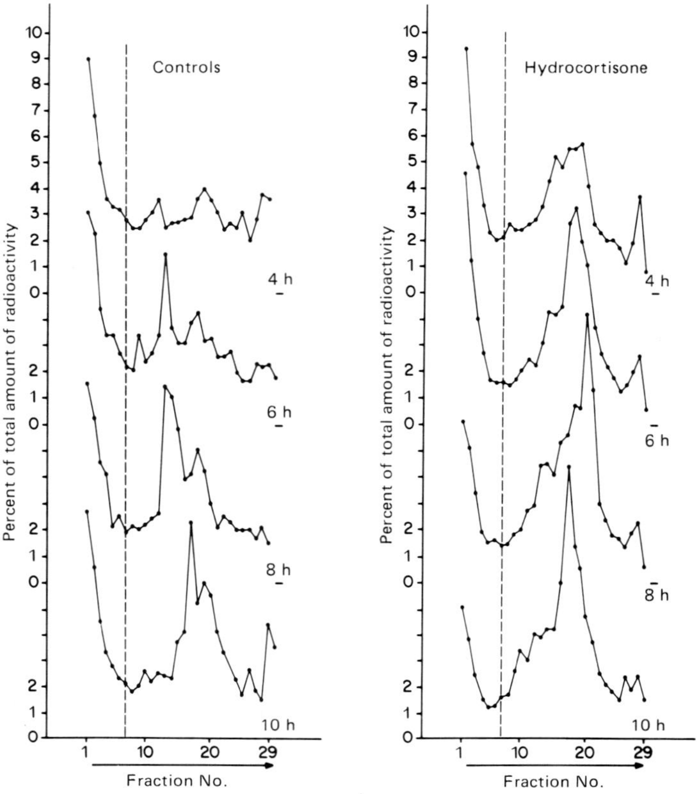

Fig. 66. Radioactivity found in the different fractions collected from alkaline sucrose gradients with chromosomal DNA from cells in phase III carried without and with hydrocortisone (5 μg/ml) during the 10 preceding population doublings. The cultures were lysed at the indicated times after adding ^{3}H-thymidine. The dashed line indicates the fraction where denaturated lambda phage DNA (40.5 S) sedimented. The sedimentation coefficient is from left to right as indicated by the arrow [Icard-Liepkalns and Macieira-Coelho, 1982].

Since hydrocortisone stimulates the growth of human embryonic fibroblasts an increases their survival [Macieira-Coelho and Loria, 1974, 1976], the sedimentation velocity of newly synthesized DNA was analysed under the same experimental conditions described above. It was found that the hormone could protect the chromatin of old cells in G_2, since DNA from hydrocortisone treated cells sedimented in a single peak after an S period (fig. 66).

The thermo- and alkali-lability of the DNA from old cell populations in the G_2 period can be explained by a defect of the gap-filling step during the rejoining of adjacent replicons due to deficient synthesis of chromosomal proteins [Ryan and Cristofalo, 1972; Mitsui et al., 1980]. On the other hand, hydrocortisone which stimulates the synthesis of DNA-bound proteins in old cells [Choe and Rose, 1974] and the acetylation and phosphorylation of histones and nonhistone chromosomal proteins [Kanungo, 1980] might this way stabilise the links between adjacent replicons.

It is interesting that in plants, extrachromosomal DNA replicates in cells that differentiate from G_2. It has been suggested that this is due to the failure of nascent replicons to join when cells reach G_2, leaving gaps that serve as recognition sites for the initiation of DNA amplification [Van't Hof and Bjerknes, 1982]. Hence, this defect in G_2 in the old cell could be an explanation for the appearance of extrachromosomal circular DNA.

d. Mitosis

The time for mitosis does not seem to change during most of the human fibroblast population life span, but at the very end abnormal mitoses can be detected [Harris, 1957; Sax and Passano, 1961; Macieira-Coelho et al., 1966a; Miller et al., 1977]. Matsumura et al. [1979b] found that towards the end of the cell life span in vitro there is an increase in multinucleated cells which is not the result of simple fission of the nucleus, but is accompanied by DNA synthesis and polyploidization. In chicken fibroblasts there is an increase in polyploidized nuclei (fig. 13, 14) and it was suggested that they result from failure of cytokinesis followed by nuclear fusion [Kaji and Matsuo, 1981]. The implication of these events for the terminal differentiation concept are discussed in chapter V.2.

Sisken and Bonner [1979] measured the time of the mitotic phases by time lapse cinematography and could not detect any changes in metaphase or anaphase. They detected, though, a 20% increase in the duration of cytokinesis at the very end of the cells' life span.

Raes et al. [1984] also observed mitoses with time lapse cinematography in hamster fibroblast cultures. They described two morphologically distinct cell types in these cultures. Type 1 cells, predominant in young populations, undergo generally normal mitosis in 60–20 min. Mitoses from type 2 cells were more heterogeneous. Some type 2 cells divide normally, some display completely aberrant mitoses, and, in others, mitosis aborts after the cell enters metaphase. In the latter case, the cell respreads into a large typical type 2 cell with an extremely polylobulated nucleus. Occasionally, these investigators could find aborted mitoses in type 1 cells. The cells seem to start cytokinesis but finally only one large non-fibroblastic cell with 2 nuclei results. They also observed asymmetric mitoses, giving rise to a binucleate cell and a cytoplasmic fragment free of nucleus. These authors suggested that these large binucleated cells resulting from aborted mitoses of type 1 cells are the intermediate stage between types 1 and 2.

Loss of chromosomal material and other abnormal mitoses were also described by Macieira-Coelho et al. [1982] in human fibroblasts and are certainly one of the causes of the genome reorganization that takes place during cell division (fig. 23).

VII. Heterogeneity of Fibroblast-Like Populations

Since most cytogerontology studies have been done with cultivated fibroblasts, it is important to stress, before concluding, some features of these cell populations which are apparent from what has already been said herein.

When one starts a culture from tissues of different vertebrates, either by protease digestion or by letting the cells migrate from an explant, the cells that rapidly invade the culture vessel have, when confluent, an elongated shape and dispose themselves in a parallel fiber-like array. It is these cells that are called fibroblasts. The fact that they overgrow so easily other cell types is due to their high mobility; a property that is needed in normal wound healing where the fibroblast rapidly invades the gap, to deposit a matrix necessary for the healing to proceed. So this high motility is in itself a marker, necessary for this important function that the fibroblast performs in vivo. The morphological counterpart of this functional aspect of the fibroblast, is the presence of 'stress fibers' visible with phase contrast polarized light or Nomarski optics, which correspond to the localization of actin and myosin by immunofluorescence [Gabbiani, 1979].

An additional marker of the normal fibroblast is the absence of the alpha-isoform of actin that can be identified by isoelectric focusing [Skalli et al., 1986], which is present in smooth muscle cells and pericytes. However, fibroblasts from the stroma of tumors and from other pathologic situations present the alpha-isoform of actin [Gabbiani, personal commun.].

The fibroblast in situ differs from the cultivated fibroblast, because it does not contain stress fibers and does not show immunofluorescent staining with antiactin, antimyosin or antitubulin antibodies [Gabbiani, 1979]. Staining occurs only when a fibrillar system develops with the differentiation of the fibroblast into myofibroblast during the formation of granulation tissue. Thus, the cultivated fibroblast has characteristics similar to the myofibroblast [Gabbiani, 1981]. Another structure present in the cultivated fibroblast and the myofibroblast of granulation tissue but not in the fibroblast in situ is the gap junction.

Granulation tissue fibroblasts may develop from local fibroblasts or from less-differentiated mesenchymal cells such as pericytes [Gabbiani, 1979]. Franks and Cooper [1972] proposed, based on morphologic studies, that the fibroblast-like cells that proliferate in vitro, are derived from pericytes. They described type 1 cells with rounded or bean-shaped nuclei, thin marginal chromatin and few organelles, type 2 cells with a convoluted nucleus, thicker marginal chromatin, prominent Golgi and endoplasmic reticulum, atypical cells suggestive of transition forms, and also some giant cells. This classification agrees with that proposed by other investigators, described in another chapter.

Cells obtained from human embryonic tissues, though not from the whole embryo, present in general the pattern of fiber-like cells in parallel arrays forming whorls with a typical design. The general pattern of cells from human postnatal skin, however, can differ quite significantly depending on the donor and may be also upon the site. Is this due to different developmental stages, postnatal cells being at a more advanced step of development which evolves according to site, age and other physiopathologic characteristics of the donor?

Whatever the answer, this morphological aspect of the cultures is an indication of their heterogeneity. As already mentioned herein, the proliferation kinetics as well as metabolic studies showed that each fibroblast population is a composite of cells with different properties. They also evolve permanently up to the end of their life span and this drift has been followed with biochemical markers. Zavala et al. [1978], for instance, followed 29 skin fibroblast cultures from 29 females heterozygous at the X-linked G6PD locus (type A or B enzyme). At the start, each of 29 cultures had double-enzyme phenotype; 10 cultures kept the double-enzyme phenotype up to the end, and 19 developed single-enzyme phenotype. The latter occurred as early as the 4th and as late as the 38th transfer.

But heterogeneity exists not only within a population, it exists also with cultures obtained from the same tissue, from different tissues, from different sites of the same tissue in the case of the skin, and also from different developmental stages. In regard to the latter aspect, cells may differ for instance when taken from embryos of different gestational stages. One example has already been given with mouse cells when discussing the problem of DNA repair. But there are other examples. Skin fibroblasts from human embryos have higher collagenase concentrations when originating from 29 weeks of gestation than from 7–8 weeks [Bauer

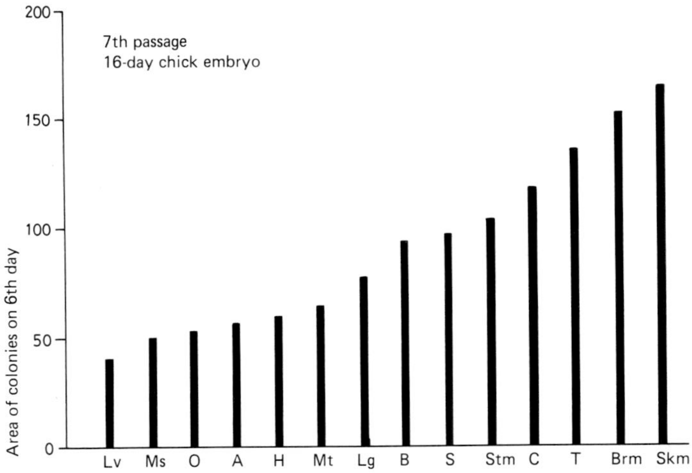

Fig. 67. Diagram representing the rate of multiplication of fourteen strains of fibroblasts isolated from the liver (Lv), mesonephros (Ms), ovary (O), aorta (A), heart (H), metanephros (Mt), lung (Lg), bone (B), spleen (S), stomach musculature (Stm), cartilage (C), thyroid (T), breast musculature (Brm), and skeletal musculature (Skm), of a 16-day chick embryo, and cultivated in a mixture of plasma and embryonic tissue juice; from planimetric measurements of the areas attained by the colonies on the 6th day of the seventh passage [Parker, 1933 by copyright permission of the Rockefeller University Press].

et al., 1985]. The response to growth factors also evolves; fibroblasts from skeletal muscle of human fetus under 50 g body weight were stimulated by TGF-beta, while tose of 100 g or greater body weight were inhibited [Hill et al., 1986]. This work is representative of the role growth factors may have during embryonic development and clearly shows that cells in vitro express aspects of the developmental stages they go through in the organism.

An early study of Parker [1933] is also revealing of the heterogeneity of fibroblast populations, or fibroblast races, as he called it, during embryonic development. The rate of multiplication of fibroblasts from different tissues of chicken embryos varied widely (fig. 67). Pertinent to the problem of the relationship between the behavior of cells in vitro and in vivo is the finding described in the same work that the cells derived from the metanephros on the 16th day, when the permanent kidney is undergoing rapid development, proliferated more actively in culture than those from the degenerating mesanephros.

More recently, Schneider et al. [1977] reported that human fetal lung cultures had faster replication rates than skin-derived cells (fig. 2). The former had also smaller cell volumes and decreased cellular RNA and protein contents. Among postnatal cells, cultures derived from the skin achieved the greatest number of doublings and those from the bone marrow yielded the least, with skeletal muscle giving intermediate response [Martin et al., 1970].

The response to corticosteroides has also been used as a marker of distinctive properties. Cell cultures established from fetal and adult human lung have a high level of cortisone-reducing activity when compared to skin cells isolated from the same individual [Smith et al., 1973; Aronson et al., 1983]. On the other hand, the stimulatory activity of hydrocortisone on human fetal lung fibroblasts [Macieira-Coelho and Loria, 1974] seems to depend on the donor [Schneider et al., 1977; Kondo et al., 1985].

Postnatal fibroblasts from different organs also differ in other respects; this is to be expected if one views the mesenchymal tissue not only as a conditioner of the microenvironment but certainly also as a reservoir for the synthesis of many specialized products. An example of the distinction between mesenchymal postnatal tissues is given by the observation of LeMonnier et al. [1980] that skin fibroblasts from the abdomen have a faster doubling time, a smaller volume, a lower glucose consumption and do not utilize serine as compared to aponeurosis fibroblasts. Fibroblasts can also differ according to the sex of the donor; it was shown by Dacis et al. [1982] that the activity of a lysosomal enzyme in cells from female donors is 30% lower than that found in male cells. This is probably due to a gene dosage effect related to the inactivation of the X-chromosome.

Some of the distinctions between the different 'races' of fibroblasts are probably necessary to provide for the metabolism of the respective organs and must play a role in the interaction with particular cell types at the sites where they are located. This is exemplified in the finding that peritoneal and synovial fibroblasts synthesize an antagonist of the C5a peptide which is chemotactic for neutrophiles [Matzner et al., 1986]. Skin fibroblasts, however, do not secrete the antagonist. This finding is consistent with the hypothesis that the fibroblasts of those cavities are regulators of inflammation at sites lined with an epithelium of mesodermal origin [Matzner et al., 1986].

Another interesting example of heterogeneity is that according to skin body site within the same individual. It has been shown by different groups that testosterone metabolism, the amount of androgen receptor and the 5-alpha-reductase activity are higher in fibroblast cultures originated from genital skin, than in those from other areas [Kaufman et al., 1975; Lamberigts et al., 1979; Griffin et al., 1981]. The skin is indeed a complex organ which has special functions according to body areas and thus it is not surprising that its fibroblasts follow that pattern.

The heterogeneity of fibroblasts is also manifested in the different response to carcinogens either because of some particularities of the donor (see chapter II) or because of their developmental stage. In this regard it has been proposed that normal embryo and newborn forskin cells are genetically unstable, which renders them more susceptible to transformation than postnatal cells (5–65 years) [Kopelovitch, 1982].

The discovery of these different properties of fibroblasts and of their role in physiology and physiopathology has been another windfall from the use of these cells in aging research.

One should add that heterogeneity within cell populations is not exclusive of fibroblasts; other types of cells, e.g. cartilage, epithelial, endothelial, smooth muscle, although not so extensively studied yet, must display the same diversity as judged from the data already available. But like fibroblasts, they maintain, in vitro, fundamental properties characteristic of their cell system.

Finally, another important aspect of fibroblast heterogeneity, which has already been mentioned, is that related with physiopathology. Bordin et al. [1984] found that fibroblast populations from human gingiva contain cells that are stimulated by the first component of complement (C1). These cells grow faster than the parent culture and than the other cells with a low affinity to C1. They also synthesize higher proportions of collagen types III and V and 40% of their protein synthesis consists in collagen production. These authors suggested that this fibroblast compartment is the one recruited in wounds and inflamed tissues. This would agree with the proposal that the composition of normal or diseased connective tissue, a consequence of a mixture of fibroblast subtypes, can be modified through deletion or amplification of the subtypes by the action of different ligands [Bordin et al., 1984]. This mechanism would play a role in the physiology and pathology of the tissues. Thus, during wound healing, the first component of complement (C1) may be a key factor recruiting the competent fibroblast compartment [Bordin et al., 1984]. An identical type of recruitment, but which would represent a deviation from normalcy has been proposed in scleroderma, where fibroblasts would be attracted by a lymphocyte-derived chemotactic factor, to sites where they are stimulated to synthesize collagen by the action of a lymphokine. This collagen triggers cell-mediated immunity for unknown reasons, so that the newly synthesized collagen may function as additional antigen to perpetrate the cycle [Kang, 1978].

Maybe some readers, puzzled by this diversity of cultivated cells, ponder if cell cultures can at all be used to study the physiopathology of the organism. However, the heterogeneity at all levels, apparent from these examples, is not surprising since diversity is a characteristic of nature; without it, probably life itself would not be possible. Diversity is the manifestation of the capacity to adapt and evolve. It is complementarity that gives the best chances of survival at the individual as well as at the cellular level; it allows metabolic processes to go on in the organism even in the presence of a failure of a large section of its elements. Complementarity and redundancy offer the potential for replacement.

So this diversity manifested in vitro is analogous to that taking place in the organism, the understanding that it exists has been another important outcome from the cultivation of cells in vitro.

It should not be forgotten, though, that in spite of the heterogeneity, normal cells in culture, as mentioned above, keep many of their markers and behave in a reproducible way, characteristic of the tissue, provided they are maintained in an appropriate way. Furthermore, there is an overwhelming evidence showing that cells in culture reproduce many aspects of the donor's physiopathology, which confirms that they maintain characteristics of the organism from where they derived.

In conclusion one can say that, fortunately, the heterogeneity is manifested when cells are explanted in vitro, it shows that tissue culture techniques reproduce what happens in the organism. It is obvious, however, that these features have to be taken into account in the experimental work. Some investigators, for instance, attributed metabolic changes to a disease when they were just the attribute of different body sites. In terms of studying the loss of the division potential, the basic mechanisms are probably identical for the different types of fibroblasts. In relation to other events, however, one should not expect that all fibroblasts in culture follow the same pattern, precisely because of their role in the respective tissue. Cells in culture can be extremely useful to study physiopathology, but, as with any experimental system, it depends upon the investigator to use them in the right way and avoid artefacts.

VIII. Envoi

Although many experiments are still unconclusive, the implications and the contributions of the research area described herein are so vast that it would be unwise to sum them up.

The use of cell culture to study aging has improved our knowledge about what happens to a normal cell during division and how it can be modified during this process. It has shown that the distribution of DNA between sister cells is asymmetric. The notion of a remodeling going on at the cellular level through division and leading to aging, which surfaced from these experiments can also be useful to explain senescence at other levels.

The analysis of proliferating cell compartments has, too, introduced the notion of uncertainty to the probability of the dividing and nondividing state. It has also provided new insight about the different phenomena regulating growth and how they relate to other cell metabolic events.

Furthermore, the investigations described herein have brought into new light several aspects of the physiopathology of the organism and revealed evolutionary differences between species. They have widened considerably the possibilities for cell culture to be used in the study of development, disease and aging.

Another contribution of this field of research is to have emphasized the importance of topological constraints for the regulation of cell behavior. In this regard, the problems raised by the mathematics of tridimensional manifolds [Thurston and Weeks, 1984] have striking implications for the cell systems described herein. It is interesting how such an event as the increase in cell volume, first described by Simons [1967] and more or less considered, when it was reported, as a descriptive curiosity, can explain many of the changes leading to the growth decline. Not the least, it can explain disturbances in intracellular communication, in particular the information flowing from the cytoplasmic membrane to the genome triggering the division cycle [Macieira-Coelho, 1983]. This has been repeatedly emphasized through this monograph. I remember a meeting where, after Dr. Simon's presentation showing an enlargement of peripheral blood cells from aged human donors, a very distinguished gerontologist raised and commented with a contempted smile, that the finding was irrelevant since these donors do very well on vitamin B12.

As the reader will certainly realise, an important aspect rendered apparent by the cell culture model for the study of aging, is that one should not expect a simple solution to explain cell senescence; let alone for the mechanisms involved in the central idea that started this long march, i.e. that somatic cells have a limited division potential which is related to aging of the organism.

I have met investigators looking for *The* event that starts the growth decline of a dividing cell compartment. This is a naive approach. I would say that the first event starting the growth decline of a dividing cell compartment occurs at the time of fertilization of the egg.

As with all new approaches, it took a long time for cell and tissue culture to be accepted for the study of aging, in spite of the previous contributions of this methodology to in vivo studies. This is probably due to the complexity of the biology of aging, which has lagged in regard to other fields of human biology. Thanks to the pioneering work of those that tried to understand the problem, the way is now open to new and rapid advances in this challenging field.

IX. References

Absher, P.M.; Absher, R.G.: Clonal variation and aging of diploid fibroblasts. Cinematographic studies of cell pedigrees. Expl Cell Res. *103:* 247–255 (1976).

Absher, P.W.; Absher, R.G.; Barnes, W.D.: Genealogies of clones of diploid fibroblasts. Cinematographic observations of cell division patterns in relation to population age. Expl Cell Res. *88:* 95–104 (1974).

Adolphe, M.; Ronot, X.; Jaffray, P.; Hecquet, C.; Fontagne, J.; Lechat, P.: Effects of donor's age on growth kinetics of rabbit articular chondrocytes in culture. Mech. Age. Dev. *23:* 191–198 (1983).

Aizawa, S.; Kurimoto, F.: Effect of glutaraldehyde and other drugs on concanavelin A-mediated red blood cell adsorption to nonsenescent, senescent and transformed human fibroblasts. Mech. Age. Dev. *11:* 237–243 (1979a).

Aizawa, S.; Kurimoto, F.: Age related cell surface changes in human diploid fibroblasts revealed by lectin-mediated red blood cell adsorption assay. A lectin survey. Mech. Age. Dev. *11:* 245–252 (1979b).

Aizawa, S.; Mitsui, Y.: A new cell surface marker of aging in human diploid fibroblasts. J. cell. Physiol. *100:* 383–387 (1979).

Aizawa, S.; Mitsui, Y.: Cell surface and clonal proliferation property of aging human diploid fibroblasts. Expl Cell Res. *139:* 416–419 (1982).

Aizawa, S.; Mitsui, Y.; Kurimoto, F.; Matsuoka, K.: Cell surface changes accompanying aging in human diploid fibroblasts. III. Division age and senescence revealed by concanavelin A-mediated red blood cell adsorption. Expl Cell Res. *125:* 297–303 (1980a).

Aizawa, S.; Mitsui, Y.; Kurimoto, F.; Matsuoka, K.: Cell-surface changes accompanying aging in diploid fibroblasts. Effect of tissue, donor age and genotype. Mech. Age. Dev. *13:* 297–306 (1980b).

Allen, S.L.; Ervin, P.R.; Mclaren, N.C.; Brand, R.E.: The 5S ribosomal RNA gene cluster in *Tetrahymena termophila.* Strain differences, chromosomal localization and loss during micronuclear aging. Mol. gen. Genet. *197:* 244–253 (1984).

Anderson, P.J.: Actin in young and senescent fibroblasts. Biochem. J. *169:* 169–172 (1978).

Aronson, J.F.; Mcclasky, J.W.; Cristofalo, V.J.: Human fetal lung fibroblasts. Observations on origin and stability in culture. Mech. Age. Dev. *21:* 229–244 (1983).

Aubery, M.; Guillouzo, A.; Bernard, B.; Font, J.: Changes in ricinus communis lectin binding to the cell surface of human liver cells in culture. Expl Cell Res. *129:* 273–280 (1980).

Asselineau, D.; Bell, E.: Prolylhydroxylase activation in IMR-90 fibroblasts; state of differentiation rather than population doubling level determines requirements for enzyme activation. Mech. Age. Dev. *25:* 257–266 (1984).

Azzarone, B.; Macieira-Coelho, A.: Heterogeneity of the kinetics of proliferation within human skin fibroblastic cell populations. J. Cell Sci. *57:* 177–187 (1982).

Azzarone, B.; Macieira-Coelho, A.: Role of cytoskeletal elements in the retractile activity of human skin fibroblasts. Expl Cell Res. *155:* 299–304 (1984).

Azzarone, B.; Diatloff-Zito, C.; Billard, C.; Macieira-Coelho, A.: Effect of low dose rate irradiation on the division potential of cells in vitro. VII. Human fibroblasts from young and adult donors. In Vitro *16:* 634–638 (1980).

Azzarone, B.; Mareel, M.; Billard, C.; Scemama, P.; Chaponnier, C.; Macieira-Coelho, A.: Abnormal properties of skin fibroblasts from cancer patients with breast cancer. Int. J. Cancer *33:* 759–764 (1984a).

Azzarone, B.; Suarez, H.; Mingari, M.C.; Moretta, L.; Fauci, A.S.: 4F2 monoclonal antibody recognizes a surface antigen on spread human fibroblasts of embryonic but not of adult origin. J. Cell Biol. *98:* 1133–1137 (1984b).

Azzencotte, R.; Courtois, Y.: Age related differences in intercellular adhesion for chicken fibroblasts cultured in vitro. Expl Cell Res. *86:* 69–74 (1974).

Bachmair, A.; Finley, D.; Varshavsky, A.: In vitro half-life of a protein is a function of its amino-terminal residue. Science *234:* 179–185 (1986).

Baker, J.B.; Humphreys, T.: Serum-stimulated release of cell contacts and the initiation of growth in contact-inhibited chick fibroblasts. Proc. natn. Acad. Sci. USA *68:* 2161–2164 (1971).

Balin, A.K.; Goodman, D.P.; Rasmussen, H.; Cristofalo, V.J.: The effect of oxygen and vitamin E on the life span of human diploid cell is vitro. J. Cell Biol. *74:* 58–67 (1977).

Ballard, F.J.; Read, L.C.: Changes in protein-synthesis and breakdown rates and responsiveness to growth-factors with aging in human lung fibroblasts. Mech. Age. Dev. *30:* 11–22 (1985).

Ban. S.; Nikaido, O.; Sugahara, T.: Modifications of doubling potential of cultured human diploid cells by ionizing radiation and hydrocortisone. Expl Geront. *15:* 539–549 (1980a).

Ban, S.; Nikaido, O.N.; Sugahara, T.: Acute and late effects of a single exposure of ionizing radiation on cultured human diploid cell populations. Radiat. Res. *81:* 120–130 (1980b).

Banks-Schlegel, S.; Green, H.: Involucrin synthesis and tissue assembly by keratinocytes in natural and cultured human epithelia. J. Cell Biol. *90:* 732–737 (1981).

Bard, J.; Elsdale, T.: Growth regulation in multilayered cultures of human diploid fibroblasts. The roles of contact, movement and matrix production. Cell Tiss. Kinet. *19:* 141–154 (1986).

Barradon, Y.; Green, H.: Cell size as a determinant of the clone-forming ability of human keratinocytes. Proc. natn. Acad. Sci. USA *82:* 5390–5394 (1985).

Bartold, P.M.; Boyd, R.R.; Page, R.C.: Proteoglycans synthesized by gingival fibroblasts derived from human donors of different ages. J. cell. Physiol. *126:* 37–46 (1986).

Basler, J.W.; David, J.D.; Agris, P.F.: Deteriorating collagen synthesis and cell ultrastructure accompanying senescence of human normal and Werner's syndrome fibroblast cell strains. Expl Cell Res. *118:* 73–84 (1979).

Bauer, E.A.; Kronberger, A.; Stricklin, G.P.; Smith, L.T.; Holbrook, K.A.: Age-related changes in collagenase expression in cultured embryonic and fetal human skin fibroblasts. Expl Cell Res. *161:* 484–494 (1985).

Bauer, E.A.; Silverman, N.; Busiek, D.F.; Kronberger, A.; Denel, T.F.: Diminished response of Werner's syndrome fibroblasts to growth factors PDGF and FGF. Science *234:* 1240–1243 (1986).

Beaupain, R.; Icard, C.; Macieira-Coelho, A.: Changes in DNA alkali-sensitive sites during senescence and establishment of fibroblasts in vitro. Biochem. biophys. Acta *606:* 251–261 (1980).

Behrman, R.G.; Wynn, V.: Low-density lipoprotein metabolism by cultured human fibroblasts. Relationship with aging and atherosclerosis. Gerontology *30:* 359–364 (1984).

Bell, E.; Marek, L.F.; Levinstone, D.S.; Merrill, C.; Sher, S.; Young, J.T.; Eden, M.: Loss of division potential in vitro. Aging or differentiation? Science *202:* 1158–1163 (1978).

Bell, E.; Marek, L.; Sher, S.; Merrill, C.; Levinstone, D.; Young, J.: Do diploid fibroblasts in culture age? Int. Rev. Cyt., suppl. 10, pp. 1–9 (1979).

Belmont, A.S. Kendall, F.M.; Nicolini, C.A.: Coupling of nuclear morphometry to cell geometry and growth in human fibroblasts. Cell Biophys. *2:* 165–175 (1980).

BeMiller, P.M.; Miller, J.E.: Cytological changes in senescing WI-38 cells. A statistical analysis. Mech. Age. Dev. *10:* 1–15 (1979).

Benn, P.A.: Specific chromosome aberrations in senescent fibroblast cell lines derived from human embryos. Am. J. hum. Genet. *28:* 465–473 (1976).

Berezney, R.; Buchholtz, L.A.: Dynamic association of replicating DNA fragments with the nuclear matrix of regenerating liver. Expl Cell Res. *132:* 1–13 (1981).

Berezney, R.; Coffey, D.S.: Nuclear matrix. Isolation and characterization of a framework structure from rat liver nuclei. J. Cell Biol. *73:* 616–637 (1977).

Berger, J.J.; Daniel, C.W.: Stromal DNA synthesis is stimulated by young, but not serially aged mouse mammary epithelium. Mech. Age. Dev. *23:* 277–284 (1983).

Bergs, V.V.; Groupe, V.: Low malignancy of Rous sarcoma cells as evidenced by poor transplantability in turkeys. Science *139:* 922–923 (1963).

Berumen, L.; Macieira-Coelho, A.: Changes in albumin uptake during the lifespan of human fibroblasts in vitro. Mech. Age. Dev. *6:* 165–172 (1977).

Beug, H.; Graf, T.: Isolation of clonal strains of chicken embryo fibroblasts. Expl Cell Res. *107:* 417–428 (1977).

Beverstock, G.C.; Simons, J.W.I.M.: Study on the variation in X-ray sensitivity of human diploid skin fibroblasts. Normal radiosensitivity of cells derived from patients with Huntington's disease. Mutation Res. *96:* 75–87 (1982).

Beyth, R.J.; Culp, L.A.: Glycosaminoglycan distribution in substratum adhesion sites of aging human skin fibroblasts including papillary and reticular subpopulations. Mech. Age. Dev. *29:* 151–169 (1985).

Bhargava, G.; Rifas, L.; Makman, M.H.: Presence of epidermal growth factor receptors and influence of epidermal growth factor on proliferation and aging in cultured smooth muscle cells. J. cell. Physiol. *100:* 365–374 (1979).

Bierman, E.L.: The effect of donor age on the in vitro life span of cultured human arterial smooth-muscle cells. In Vitro *14:* 951–955 (1978).

Bierman, E.L.; Albers, J.J.; Chait, A.: Effect of donor age on the binding and degradation of low density lipoproteins by cultured human arterial smooth muscle cells. J. Geront. *34:* 483–488 (1979).

Bierman, E.L.; Schwartz, S.M.: Effect of clonal senescence on low density lipoprotein-receptor activity of bovine arterial endothelial cells. In Vitro *20:* 809–813 (1984).

Bittles, A.H.; Harper, N.: Increased glycolysis in aging cultured human diploid fibroblasts. Biosci. Rep. *4:* 751–756 (1984).

Blomqvist, E.; Arro, E.; Brunk, U.; Westermark, B.: Plasma membrane motility of cultured human glia cells in phase II and III. Acta pathol. microbiol. scand., A, Pathol. *86:* 257–263 (1978).

Blomqvist, E.; Macieira-Coelho, A.; Brunk, U.: The influence of cell cooperation, nutrients and surface area on cell division (submitted for publ.).

Blondal, J.A.; Dick, J.E.; Wright, J.A.: Membrane glycoprotein changes during the senescence of normal human diploid fibroblasts in culture. Mech. Age. Dev. *30:* 273–283 (1985).

Boak, A.M.; Bittles, A.H.; Quinn, P.J.: Age related ultrastructural changes in human embryonic lung fibroblasts. Expl. Gerontol. *18:* 139–146 (1983).

Bordin, S.; Page, R.C.; Narayanan, A.S.: Heterogeneity of normal human diploid fibroblasts: isolation and characterization of one phenotype. Science *223:* 171–172 (1984).

Bosmann, H.B.; Gutheil, R.L., Jr.; Case, K.R.: Loss of a critical neutral protease in ageing WI-38 cells. Nature Lond. *261:* 499–501 (1976).

Bourgeois, C.A.; Raynaud, N.; Diatloff-Zito, C.; Macieira-Coelho, A.: Effect of low dose rate ionizing radiatrion on the division potential of cells in vitro. VIII. Cytogenetic analysis of human fibroblasts. Mech. Age. Dev. *17:* 225–235 (1981).

Bowman, P.D.; Daniel, C.W.: Aging of human fibroblasts in vitro, surface features and behaviour of aging WI-38 cells. Mech. Age. Dev. *4:* 147–158 (1975).

Bowman, P.D.; Meek, R.L.; Daniel, C.W.: Aging of human fibroblasts in vitro, correlation between DNA synthetic ability and cell size. Expl Cell Res. *93:* 184–190 (1975).

Bowman, P.D.; Meek, R.L.; Daniel, C.W.: Decreased unsheduled DNA synthesis in nondividing aged WI-38 cells. Mech. Age. Dev. *5:* 251–257 (1976a).

Bowman, P.D.; Meek, R.L.; Daniel, C.W.: Decreased synthesis of nuclear RNA in aging human cells in vitro. Expl Cell Res. *101:* 434–437 (1976b).

Brash, K.: Fine structure and localization of the nuclear matrix in situ. Expl Cell Res. *140:* 161–171 (1982).

Bradley, M.O.; Erikson, L.C.; Kohn, K.W.: Normal DNA strand rejoining and absence of DNA crosslinking in progeroid and aging human cells. Mutation Res. *37:* 279–292 (1976a).

Bradley, M.O.; Dice, J.F.; Hayflick, L.; Schimke, R.T.: Protein alterations in aging WI-38 cells as determined by proteolytic susceptibility. Expl Cell Res. *96:* 103–112 (1975).

Bradley, M.O.; Hayflick, L.; Schimke, R.T.: Protein degradation in human fibroblasts (WI-38), effect of aging, viral transformation and amino acid analogs. J. biol. Chem. *251:* 3521–3529 (1976b).

Brandes, D.; Murphy, D.G.; Anton, E.B.; Barnard, S.: Ultrastructural and cytochemical changes in cultured human lung cells. J. ultrastruct. Res. *39:* 465–483 (1972).

Breul, S.D.; Bradley, K.H.; Hance, A.J.; Schafer, M.P.; Berg, R.A.; Crystal, R.G.: Control of collagen production by human diploid lung fibroblasts. J. biol. Chem. *255:* 5250–5260 (1980).

Brissenden, J.E.; Cox, D.W.: Alpha-2-macroglobulin production by cultured human fibroblasts. Som. Cell Genet. *8:* 289–305 (1982).

Brock, M.A.; Hay, R.J.: Comparative ultrastructure of chick fibroblasts in vitro at early and late stages during their growth span. J. ultrastruct. Res. *36:* 291–302 (1971).

Brown, W.T.; Ford, J.P.; Gershey, E.L.: Variation of DNA repair capacity in Progeria cells unrelated to growth conditions. Biochem. Biophys. Res. Commun. *97:* 347–353 (1980a).

Brown, W.T.; Darlington, G.J.: Thermolabile enzymes in Progeria and Werner syndrome. Evidence contrary to the protein error hypothesis. Am. J. hum. Genet. *32:* 614–619 (1980b).

Brown, W.T.; Zebrower, M.; Kieras, F.J.: Progeria, a model disease for the study of accelerated aging; in Woodhead, Blackett, Hollaender, Molecular biology of aging, pp. 375–396 (Plenum Press, New York 1985).

Bruce, S.A.; Deamond, S.F.; Ts'o, P.O.P.: In vitro senescence of Syrian hamster mesenchymal cells of fetal to aged adult origin. Inverse relationship between in vivo donor age and in vitro proliferative capacity. Mech. Age. Dev. *34:* 151–173 (1986).

Brunk, U.: Distribution and shift of ingested marker particles in residual bodies and other lysosomes. Studies on in vitro cultivated human glia cells in phase II and III. Expl Cell Res. *79:* 15–27 (1973).

Brunk, U.; Ericson, J.L.E.; Pontén, J.; Westermark, B.: Residual bodies and 'aging' in cultured human glia cells. Effect of entrance into phase III and prolonged periods of confluence. Expl Cell Res. *79:* 1–14 (1973).

Brunk, U.; Schellens, J.; Westermark, B.: Influence of epidermal growth factor (EGF) on ruffling activity, pinocytosis and proliferation of cultivated human glia cells. Expl Cell Res. *103:* 295–302 (1976).

Buchanan, J.H.; Bunn, C.L.; Lappin, R.J.; Stevens, A.: Accuracy of in vitro protein synthesis. Translation of polyuridic acid by cell-free extracts of human fibroblasts. Mech. Age. Dev. *12:* 339–353 (1980).

Buchanan, J.H.; Sidhu, J.: Autofluorescence and ageing. Changes in ribosome accuracy and lysosome function. Mech. Age. Dev. *36:* 259–267 (1986).

Buchanan, J.H.; Stevens, A.: Fidelity of histone synthesis in cultured human fibroblasts. Mech. Age. Dev. *7:* 321–334 (1978).

Bunn, C.L.; Tarrant, G.M.: Limited life span in somatic cell hybrids and cybrids. Expl Cell Res. *127:* 385–396 (1980).

Burmer, G.C.; Motulsky, H.; Zeigler, C.J.; Norwood, T.H.: Inhibition of DNA synthesis in young cycling human diploid fibroblast-like cell upon fusion to enucleate cytoplasts from senescent cells. Expl Cell Res. *145:* 79–84 (1983).

Burmer, G.C.; Norwood, T.H.: Selective elimination of proliferating cells in human diploid cell cultures by treatment with BrdU, 33258 Hoechst and visible light. Mech. Age. Dev. *12:* 151–159 (1980).

Burmer, G.C.; Zeigler, C.J.; Norwood, T.H.: Evidence for endogenous polypeptide-mediated inhibition of cell-cycle transit in human diploid cells. J. Cell Biol. *94:* 187–192 (1982).

Burnet, M.: Endurance of life, p. 19 (Cambridge University Press, London 1978).

Burns, F.J.; Tannock, J.F.: On the existence of a Go-phase in the cell cycle. Cell Tiss. Kinet. *3:* 321–325 (1970).

Cameron, I.L.: Minimum number of cell doublings in an epithelial cell population during the life-span of the mouse. J. Geront. *27:* 157–161 (1972).

Capanna, E.: Glotaxonomy and vertebrate evolution, p. 783 (Academic Press, London 1973).

Carlin, C.R.; Phillips, P.D.; Knowles, B.B.; Cristofalo, V.J.: Diminished in vitro tyrosine kinase activity of the EGF receptor of senescent human fibroblasts. Nature, Lond. *306:* 617–620 (1983a).

Carlin, C.; Golczewski, J.; Phillips, P.D.; Knowles, B.; Cristofalo, V.J.: Senescent WI-38 cells bind epidermal growth factor but fail to express the EGF receptor-associated protein kinase activity. In Vitro *19:* 282A (1983b).

Carnicero, H.H.; Adamany, A.M.; Englard, S.: Collagen glucosyl- and galactosyl transferases of cultured human fetal lung fibroblasts. Archs. Biochem. Biophys. *210:* 678–690 (1981).

Carrel, A.; Ebeling, A.H.: Age and multiplication of fibroblasts. J. exp. Med. *34:* 599–623 (1921).

Cassiman, J.J.; David, G.; Schueren, B. van der; Berghe, H. van den: Differences in agglutinability of adult and fetal human fibroblasts using phytohemagglutinin. Cell Differ. *6:* 133–145 (1977).

Castor, L.N.: A G1 rate model accounts for cell-cycle kinetics attributed to 'transition probability'. Nature, Lond. *287:* 857–859 (1980).

Cavallius, J.; Rattan, S.I.S.; Clark, B.F.C.: Changes in activity and amount of active elongation factor 1 alpha in aging and immortal human fibroblast cultures. Expl. Geront. *21:* 149–157 (1986).

Celis, J.E.; Bravo, R.: Synthesis of the nuclear protein cyclin in growing, senescent and morphologically transformed human skin fibroblasts. FEBS Lett. *165:* 21–25 (1984).

Chan, A.C.; Walker, J.G.: Loss of DNA repair capacity during successive subcultures of primary rat fibroblasts. J. Cell Biol. *74:* 365–370 (1977).

Chandrasekhar, S.; Norton, E.; Millis, A.J.T.; Izzard, C.S.: Functional changes in cellular fibronectin from late passage fibroblasts in vitro. Cell Biol. int. Rep. *7:* 11–21 (1983a).

Chandrasekhar, S.; Sorrentino, J.A.; Millis, A.J.T.: Interaction of fibronectin with collagen. Age-specific defect in the biological activity of human fibroblasts fibronectin. Proc. natn. Acad. Sci. USA *80:* 4747–4751 (1983b).

Chen, T.R.; Ruddle, F.H.: Chromosome changes revealed by the Q-band staining method during cell senescence of WI-38. Proc. Soc. exp. Biol. Med. *147:* 533–536 (1974).

Chapman, M.L.; Zaun, M.R.; Gracy, R.W.: Changes in NAD levels in human lymphocytes and fibroblasts during aging and in premature aging syndromes. Mech. Age. Dev. *21:* 157–167 (1983).

Chen, J.J.; Brot, N.; Weissbach, H.: RNA and protein synthesis in cultured human fibroblasts derived from donors of various ages. Mech. Age. Dev. *13:* 285–296 (1980).

Chen, K.Y.; Chang, Z.F.: Age dependency of the metabolic conversion of polyamines into amino acids in IMR-90 human embryonic lung diploid fibroblasts. J. cell. Physiol. *128:* 27–32 (1986).

Chen, K.Y.; Chang, Z.F.; Liu, A.Y.C.: Changes of serum-induced ornithine decarboxylase during aging of IMR-90 human diploid fibroblasts. J. cell. Physiol. *129:* 142–146 (1986).

Choe, B.K.; Rose, N.R.: Synthesis of DNA binding proteins during the cell cycle of WI-38 cells. Expl Cell Res. *83:* 271–280 (1974).

Chua, C.C.; Geiman, D.E.; Ladda, R.L.: Receptor for epidermal growth factor retains normal structure and function in aging cells. Mech. Age. Dev. *34:* 35–55 (1986).

Clarkson, J.M.N.; Painter, R.B.: Repair of X-Ray damage in aging WI-38 cells. Mutation Res. *23:* 107–112 (1974).

Cleaver, J.E.: Defective repair replication of DNA in xeroderma pigmentosum. Nature, Lond. *218:* 652–656 (1968).

Cleaver, J.E.: Xeroderma pigmentosum. A human disease in which an initial stage of DNA repair is defective. Proc. natn. Acad. Sci. USA *63:* 428–435 (1969).

Cleaver, J.E.: DNA repair and its coupling to DNA replication in eukaryotic cells. Biochim. biophys. Acta *516:* 489–516 (1978).

Cleaver, J.E.: DNA repair deficiencies and cellular senescence are unrelated in xeroderma pigmentosum cell lines. Mech. Age. Dev. *27:* 189–196 (1984).

Clemmons, D.R.: Age dependent production of a competence factor by human fibroblasts. J. cell. Physiol. *114:* 61–67 (1983).

Clemmons, D.R.; Shaw, D.S.: Variables controlling somatomedin production by cultured human fibroblasts. J. cell. Physiol. *142:* 115–137 (1983).

Collins, A.R.S.; Squires, S.; Johnson, R.T.: Inhibitors of repair DNA synthesis. Nucl. Acids Res. *10:* 1203–1213 (1982).

Collins, J.M.: Deoxyribonucleic acid structure in human diploid fibroblasts stimulated to proliferate. J. biol. Chem. *252:* 141–147 (1977).

Collins, J.M.; Chu, A.K.: Reduction of DNA synthesis in aging but still proliferating cells. J. cell. Physiol. *124:* 165–174 (1985).

Collins, S.J.; Gallo, R.C.; Gallagher, R.E.: Continuous growth and differentiation of human myeloid leukaemic cells in suspension culture. Nature, Lond. *270:* 347–349 (1977).

Collins, V.P.; Arro, E.; Blomqvist, E.; Brunk, U.; Frederikson, B.A.; Westermark, B.: Cell locomotion and proliferation in relation to available surface area, serum concentration and culture age. Scan. electr. Microsc. *111:* 411–420 (1979).

Collins, V.P.; Brunk, U.: Characterization of residual bodies formed in phase II cultivated human glia cells. Mech. Age. Dev. *5:* 193–207 (1976).

Collins, V.P.; Thaw, H.H.: The measurement of lipoid peroxidation products (lipofuscin) in individual cultivated human glial cells. Mech. Age. Dev. *23:* 199–214 (1983).

Conway, H.: Differential diagnosis of keloids and hypertrophic scars by tissue culture technique, with notes on therapy of keloids by surgical excision and Decadron. Plast. reconstr. Surg. *25:* 117–122 (1960).

Cooke, A., Harris, W.J.: Pyrimidine dimer excision in aging human cells in tissue culture. Int. J. radiat. Biol. *39:* 223–226 (1981).

Cooke, A.; Harris, W.J.: Newly synthesized DNA in aging human cells in culture treated with 4-nitroquinoline-1-oxide. Mech. Age. Dev. *19:* 171–180 (1982).

Condon, M.A.A.; Oski, F.A.; Dimauro, S.; Mellman, W.J.: Glycolytic difference between foetal and non foetal human fibroblast lines. Nature New Biol. *229:* 214–215 (1971).

Conover, C.A.; Rosenfeld, R.C.; Hintz, R.L.: Hormonal control of the replication of human fetal fibroblasts. Role of somatomedin c/insulin-like growth factor I. J. cell. Physiol. *128:* 47–54 (1986).

Cooper, J.T.; Goldstein, S.: Comparative studies on human skin fibroblasts: lifespan and lipid metabolism in medium containing fetal bovine or human serum. In Vitro *13:* 473–476 (1977).

Courtois, Y.G.C.: Chromatin modification of chick embryo cells during in vitro senescence. Mech. Age. Dev. *3:* 51–63 (1974).

Cox, R.; Masson, W.K.: Changes in radiosensitivity during the in vitro growth of diploid human fibroblasts. Int. J. radiat. Biol. *26:* 193–196 (1974).

Cremer, T.; Werdan, K.; Stevenson, A.F.G.; Lehner, K.; Messerschmidt, O.: Aging in vitro and d-glucose uptake kinetics of diploid human fibroblasts. J. cell. Physiol. *106:* 99–108 (1981).

Cristofalo, V.J.: Animal cell cultures as a model system for the study of aging; in Strehler, Advances in gerontology research, vol. 4, pp. 45–79 (Academic Press, New York 1972).

Cristofalo, V.; Kabakjian, J.: Lysosomal enzymes and aging in vitro. Subcellular enzyme distribution and effect of hydrocortisone on cell life span. Mech. Age. Dev. *4:* 19–28 (1975).

Cristofalo, V.J.; Kritchevsky, D.: Respiration and glycolysis in the human diploid cell strain WI-38. J. cell. Physiol. *67:* 125–132 (1966).

Cristofalo, V.J.; Kritchevsky, D.: Cell size and nucleic acid content in the diploid human cell line WI-38 during aging. Med. Exp. *19:* 313–320 (1969).

Cristofalo, V.J.; Parris, N.; Kritchevsky, D.: Enzyme activity during the growth and aging of human cells in vitro. J. cell. Physiol. *69:* 263–272 (1967).

Cristofalo, V.J.; Sharf, B.B.: Cellular senescence and DNA synthesis. Thymidine incorporation as a measure of population age in human diploid cells. Expl Cell Res. *76:* 419–427 (1973).

Crossin, K.L.; Carney, D.H.: Evidence that microtubule depolymerization early in the cell cycle is sufficient to initiate DNA synthesis. Cell *23:* 61–71 (1981).

Crosti, N.; Bajer, J.; Serra, A.; Rigo, A.; Scarpa, M.; Viglino, P.: Coordinate expression of Mn SOD and Cu Zn SOD in human fibroblasts. Expl Cell Res. *160:* 396–402 (1985).

Croute, F.; Vidal, S.; Soleilhavoup, J.P.; Vincent, C.; Serre, G.; Planel, H.: Effects of a very low dose rate of chronic ionizing radiation on the division potential of human embryonic lung fibroblasts in vitro. Expl Geront. *21:* 1–11 (1986).

Cudkowicz, G.; Upton, A.C.; Shearer, G.M.: Lymphocyte content and proliferative capacity of serially transplanted mouse bone marrow. Nature, Lond. *201:* 165–166 (1964).

Cunha, G.R.: Effects of sex hormones on epithelial cells are mediated via trophin influences from stroma. In Vitro Cell Dev. Biol. *22:* 49A (1986).

Cutler, R.G.: The dysdifferentiation hypothesis of mammalian aging and longevity; in Giacobini, Filogamo, Vernadakis, The aging brain, vol. 20, pp. 25–114 (Raven Press, New York 1982).

Daniel, C.W.; Young, L.J.T.: Influence of cell division on an aging process. Life span of mouse mammary epithelium during serial propagation in vivo. Expl Cell Res. *65:* 27–32 (1971).

Danot, M.; Gershon, H.: The lack of altered enzyme molecules in 'senescent' mouse embryo fibroblasts in culture. Mech. Age. Dev. *4:* 289–299 (1975).

Darlington, G.J.; Dutkowski, R.; Brown, W.T.: Sister chromatid exchange frequencies in Progeria and Werner syndrome patients. Am. J. hum. Genet. *33:* 762–766 (1981).

Davidson, R.G.; Nitowsky, H.M.; Childs, B.: Demonstration of two populations of cells in the human female heterozygous for glucose-6-phosphate dehydrogenase variants. Proc. natn. Acad. Sci. USA *50:* 481–485 (1963).

Danner, D.B.; Schneider, E.L.; Pitha, J.: Macromolecular synthesis in human diploid fibroblasts. A viral probe examining the effect of in vivo aging. Expl Cell Res. *114:* 63–67 (1978).

Davies, R.L.; Fuhrer-Krusi, S.; Kucherlapati, R.: Modulation of transfected gene expression mediated by changes in chromatin structure. Cell *31:* 521–529 (1984).

Davis, M.H.; Pato, C.N.; Grinvalsky, H.; Gruenstein, E.: Identification of a biochemical difference between male and female human fibroblasts. Implications for the expression of Duchenne muscular dystrophy. Neurology, Minneap. *32:* 951–957 (1982).

Deamer, D.W.; Gonzales, J.: Autofluorescent structures in cultured WI-38 cells. Archs Biochem. Biophys. *165:* 421–426 (1974).

Dean, R.; Kim, S.S.; Delgado, D.: Expression of c-myc oncogene in human fibroblasts during in vitro senescence. Biochem. biophys. Res. Commun. *135:* 105–109 (1986).

Dean, R.T.; Riley, P.A.: The degradation of normal and analogue containing proteins in MCR-5 fibroblasts. Biochim. Biophys. Acta *539:* 230–237 (1978).

Delinassios, J.G.: Prolonged in vitro maintenance of human diploid fibroblasts in a new tissue culture medium (PRC-1). Expl Cell Biol. *51:* 315–321 (1983).

Dell'Orco, R.T.; Andersen, L.E.: Unscheduled DNA synthesis in human diploid cells of different donor ages. Cell Biol. int. Rep. *5:* 359–364 (1981).

Dell'Orco, R.T.; Andersen, L.E.; Conway, E.; Birckbichler, P.J.: Variable transglutaminase activity in human diploid fibroblasts during in vitro senescence. Cell Biol. int. Rep. *9:* 945–956 (1985).

Dell'Orco, R.T.; Andersen, L.E.; Whittle, W.L.: The effect of UV irradiation on proliferation and life span of human diploid fibroblast-like cells. In Vitro *18:* 703–707 (1982).

Dell'Orco, R.T.; Guthrie, P.L.: Altered protein metabolism in arrested populations of aging human diploid fibroblasts. Mech. Age. Dev. *5:* 399–407 (1976).

Dell'Orco, R.T.; Mertens, J.G.; and Kruse, P.F., Jr.: Doubling potential, calendar time and senescence of human diploid cells in culture. Expl Cell Res. *77:* 356–360 (1973).

Dell'Orco, R.T.; Mertens, J.G.; Kruse, P.F., Jr.: Doubling potential, calendar time and donor age of human diploid cells in culture. Expl Cell Res. *84:* 363–366 (1974).

Dell'Orco, R.T.; Whittle, W.L.: Unscheduled DNA synthesis in confluent and mitotically arrested populations of aging human diploid fibroblasts. Mech. Age. Dev. *8:* 269–279 (1978).

Dell'Orco, R.T.; Whittle, W.L.: Evidence for an increased level of DNA damage in high doubling level human diploid cell in culture. Mech. Age. Dev. *15:* 141–152 (1981).

Dell'Orco, R.T.; Whittle, W.L.: Micrococal nuclease and DNase I digestion of DNA from aging human diploid cells. Biochem. Biophys. Res. Commun. *107:* 117–122 (1982).

Dell'Orco, R.T.; Whittle, W.L.; Duncan, M.R.; Robinson, M.J.: The effects of histones, in vitro age and culture state on the digestion of DNA by micrococal nuclease and deoxyribonuclease I. Mech. Age. Dev. *21:* 273–282 (1983).

Dell'Orco, R.T.; Whittle, W.L.; Macieira-Coelho, A.: Changes in the higher order organization of DNA during aging of human fibroblast-like cells. Mech. Age. Dev. *35:* 199–208 (1986).

Diatloff, C.; Macieira-Coelho, A.: Effect of low-dose-rate irradiation on the division potential of cells in vitro. V. Human skin fibroblasts from donors with a high risk of cancer. J. natn. Cancer Inst. *63:* 55–59 (1979).

Diatloff-Zito, C.; Macieira-Coelho, A.: Effect of growth arrest on the doubling potential of human fibroblasts in vitro. A possible influence of the donor. In Vitro *18:* 606–610 (1982).

Dice, J.F.: Altered degradation of proteins micro-injected into senescent human fibroblasts. J. biol. Chem. *257:* 14624–14627 (1982).

Dick, J.E.; Wright, J.A.: Involvement of ribonucleotide reductase activity in the senescence of normal human diploid fibroblasts. Mech. Age. Dev. *20:* 103–109 (1982).

Dick, J.E.; Wright, J.A.: Ribonucleotide reductase activity during the senescence of normal human diploid fibroblasts in culture. Mech. Age. Dev. *32:* 85–97 (1985a).

Dick, J.E.; Wright, J.A.: On the importance of deoxyribonucleotide pools in the senescence of cultured human diploid fibroblasts. FEBS Lett. *179:* 21–24 (1985b).

Didinsky, J.B.; Rheinwald, J.G.: Failure of hydrocortisone or growth factors to influence the senescence of fibroblasts in a new culture system for assessing replicative lifespan. J. cell. Physiol. *109:* 171–179 (1981).

Dinowitz, M.: A continuous line of Rous sarcome virus-transformed chick embryo cells. J. natn. Cancer Inst. *58:* 307–312 (1977).

Dominice, J.; Levasseur, C.; Larno, S.; Ronot, X.; Adolphe, M.: Age-related changes in rabbit articular chondrocytes. Mech. Age. Dev. *37:* 231–240 (1986).

Drescher-Lincoln, C.K.; Smith, J.R.: Inhibition of DNA synthesis in proliferating human diploid fibroblasts by fusion with senescent cytoplasts. Expl Cell Res. *144:* 455–462 (1983).

Drescher-Lincoln, C.K.; Smith, J.R.: Inhibition of DNA synthesis in senescent-proliferating cybrids is mediated by endogenous proteins. Expl Cell Res. *153:* 208–217 (1984).

Duffy, P.E.; Kremzner, L.T.: Ornithine decarboxylase activity and polyamines in relation to aging of human fibroblasts. Expl Cell Res. *108:* 435–440 (1977).

Duncan, M.R.; Dell'Orco, R.T.: Effects of the incorporation of *p*-fluorophenylalanine on the in vitro lifespan of human diploid cells. Proc. Soc. exp. Biol. Med. *161:* 449–451 (1979).

Duncan, M.R.; Dell'Orco, R.T.; Guthrie, P.L.: Relationship of heat labile glucose 6-phosphate dehydrogenase and multiple molecular forms of the enzyme in senescent human fibroblasts. J. cell. Physiol. *93:* 49–56 (1977).

Duncan, M.R.; Dell'Orco, R.T.; Kirk, K.D.: Superoxide dismutase specific activities in cultured human diploid cells of various donor ages. J. cell. Physiol. *98:* 437–442 (1979).

Eagle, H.: Amino acid metabolism in mammalian cell cultures. Science *130:* 432–437 (1959).

Elias, J.A.; Zurier, R.B.; Schreiber, A.D.; Leff, J.A.; Daniele, R.P.: Monocyte inhibition of lung fibroblasts growth: relationship to fibroblast prostaglandin production and density defined monocyte subpopulations. J. Leuk. Biol. *37:* 15–28 (1985).

Elliot, G.C.; Johnson, R.I.: DNA repair in mouse embryo fibroblasts. II. Responses of nontransformed preneoplastic and tumorigenic cells to ultraviolet irradiation. Mutation. Res. *145:* 185–194 (1985).

Elliot, M.S.; Katze, J.R.: Inhibition of queuine uptake in diploid human fibroblasts by phorbol-12,13-didecanoate. J. biol. Chem. *261:* 13019–13025 (1986).

Engelhardt, D.E.; Lee, G.T.Y.; Moley, J.F.: Patterns of peptide synthesis in senescent and presenescent human fibroblasts. J. cell. Physiol. *98:* 193–198 (1979).

Engström-Laurent, A.; Feltelius, N.; Hällgren, R.; Wasteson, Å.: Raised serum hyaluronate levels in scleroderma: an effect of growth factor induced activation of connective tissue cells? Ann. Rheum. Dis. *44:* 614–620 (1985).

Evans, C.H.; Gansen, P. van; Rosson, I.: Transcription in early and late passage embryonic mouse fibroblasts in primary culture. A correlated biochemical, autoradiographical and ultrastructural study. Biol. Cell *33:* 117–128 (1978).

Evans, C.H.; Georgescu, H.J.: Observations on the senescence of cells derived from articular cartilage. Mech. Age. Dev. *22:* 179–191 (1983).

Fairweather, D.S.; Fox, M.; Margison, G.: The in vitro lifespan of MRC-5 cells is shortened by 5-Azacytidine-induced demethylation. Expl Cell Res. *168:* 153–159 (1987).

Feinmark, S.J.; Bailey, J.M.: Lipid metabolism in cultured cells. J. biol. Chem. *257:* 2816–2821 (1982).

Finlay, C.A.; Rosner, B.A.; Miller, J.E.; Wallace, J.H.; Cristofalo, V.J.: Enhanced proliferation of WI-38 cells in the presence of glucocorticoid-conditioned medium. Expl Cell Res. *156:* 462–470 (1985).

Fletcher, B.L.; Dillard, C.J.; Tappel, A.L.: Measurement of fluorescent lipid peroxidation products in biological systems and tissues. Analyt. Biochem. *52:* 1–6 (1973).

Folkman, J.; Moscona, A.: Role of cell shape in growth control. Nature, Lond. *273:* 345–349 (1978).

Francis, A.A.; Lee, W.H.; Regan, J.D.: The relationship of DNA excision repair of ultraviolet-induced lesions to the maximum life span of mammals. Mech. Age. Dev. *16:* 181–189 (1981).

Franks, L.M.; Cooper, T.W.: The origin of human embryo lung cells in culture. A comment on cell differentiation, in vitro growth and neoplasia. Int. J. Cancer *9:* 19–29 (1972).

Fry, M.; Weissman-Shomer, P.: Altered nuclear deoxyribonucleic acid alpha polymerases in senescent chick embryo fibroblasts. Biochemistry *15:* 4319–4329 (1976).

Fuchs, E.; Green, H.: Changes in keratin gene expression during terminal differentiation of the keratinocyte. Cell *19:* 1033–1042 (1980).

Fujiwara, Y.; Higashikawa, T.; Tatsumi, M.: A retarded rate of DNA replication and normal level of DNA repair in Werner's Syndrome fibroblasts in culture. J. cell. Physiol. *92:* 365–374 (1977).

Funfu, D.L.M.; Houbion, A.; Raes, M.; Remacle, J.: Subcellular localization and modification with aging of glutathione, glutathione peroxidase and glutathione reductase activities in human fibroblasts. Biochim. biophys. Acta *838:* 211–220 (1985).

Gabbiani, G.: The role of contractile proteins in normal healing and fibrocontractive diseases; in Jasmin, Cantin, Methods and achievements in experimental pathology, vol. 9, pp. 187–206 (Karger, Basel 1979).

Gabbiani, G.; Rungger-Brändle, E.: The fibroblast; in Glynn, Handbook of inflammation, vol. III, pp. 1–50 (Elsevier/Nord-Holland/Biomedical Press, Amsterdam 1981).

Garcia-Giralt, E.; Berumen, L.; Macieira-Coelho, A.: Growth inhibitory activity in the supernatants of nondividing WI-38 cells. J. natn. Cancer Inst. *45:* 649–655 (1970).

Gansen, P. van; Siebertz, B.; Capone, B.; Malherbe, L.: Relationship between cytoplasmic microtubular complex, DNA synthesis and cell morphology in mouse embryonic fibroblasts (effects of age, serum deprivation, aphidicolin, cytochalasin B and colchicine). Biol. Cell *52:* 161–174 (1984).

Gaubatz, J.W.; Cutler, R.G.: Age related differences in the number of ribosomal RNA genes of mouse tissues. Gerontology *24:* 179–207 (1978).

Geder, L.; Vaczi, L.; Boldogh, I.: Development of cell lines after exposure of chicken embryonic fibroblasts to herpes simplex virus type 2 at supraotimal temperature. Acta microbiol. Acad. Sci. Hung. *20:* 119–125 (1973).

Gedigk, P.; Bontke, E.: Über den Nachweis von hydrolytischen Enzymen in Lipopigmenten. Z. Zellforsch. *44:* 495–510 (1956).

Gelfant, S.; Smith, J.G.: Aging. Noncycling cells, an explanation. Science *178:* 357–361 (1972).

Germinario, R.J.; Mcquillan, A.: Glucorticoid-induced modulation of insulin-stimulated DNA synthesis. Differential responsiveness in cell cultures derived from donors of different ages. Gerontology *31:* 65–75 (1985).

Germinario, R.J.; Oliveira, M.; Taylor, M.: Studies on the effects of in vitro ageing on saturable and nonsaturable sugar uptake in cultured human skin fibroblasts. Gerontology *26:* 181–187 (1980).

Germinario, R.J.; Oliveira, M.; Manuel, S.; Taylor, M.: Characteristics of normal and maturity-onset diabetic (type II diabetes) cell cultures. Life spans and DNA synthetic capabilities. Gerontology *32:* 148–157 (1986).

Geuskens, M.; Preumont, A.M.; Gansen, P. van: Fibronectin localization and endocytosis in early and late mouse embryonic fibroblasts in primary culture. A study by light and electron microscopy immunocytochemistry. Mech. Age. Dev. *33:* 191–209 (1986).

Gilchrest, B.A.: Prior chronic sun exposure decreases the lifespan of human skin fibroblasts in vitro. J. Geront. *35:* 537–541 (1980).

Girardi, A.J.; Jensen, F.C.; Koprowski, H.: SV40-induced transformation of human diploid cells: crisis and recovery. J. cell. comp. Physiol. *65:* 69–78 (1965).

Golczewski, J.A.: Effect of 2-mercaptoethylamine on proliferation and lifespan of WI-38 cells. Expl Geront. *19:* 7–11 (1984).

Goldstein, J.L.; Brown, M.S.: Binding and degradation of low density lipoproteins by cultured human fibroblasts. J. biol. Chem. *249:* 5153–5162 (1974).

Goldstein, S.: The role of DNA repair in aging of cultured fibroblasts from xeroderma pigmentosum and normals. Proc. Soc. exp. Biol. Med. *137:* 730–734 (1971).

Goldstein, S.: Aging in vitro. Growth of cultured cells from the Galapagos tortoise. Expl Cell Res. *83:* 297–301 (1974).

Goldstein, S.; Ballantyne, S.R.; Robson, A.L.; Moerman, E.J.: Energy metabolism in cultured human fibroblasts during aging in vitro. J. cell. Physiol. *112:* 419–424 (1982).

Goldstein, S.; Csullog, G.W.: Macromolecular synthesis in human fibroblasts at 37 °C and 42 °C during aging in vitro. Mech. Age. Dev. *6:* 185–195 (1977).

Goldstein, S.; Korczak, L.B.: Status of mitochondria in living human fibroblasts during growth and senescence in vitro. Use of the laser dye rhodamine 123. J. Cell Biol. *91:* 392–398 (1981).

Goldstein, S.; Lin, C.C.: Rescue of senescent human fibroblasts by hybridization with hamster cells in vitro. Expl Cell Res. *70:* 436–439 (1971).

Goldstein, S.; Moerman, E.J.: Heat-labile enzymes in Werner's syndrome fibroblasts. Nature, Lond. *255:* 159 (1975).

Goldstein, S.; Moerman, E.J.; Soeldner, J.S.; Gleason, R.E.; Barnett, D.M.: Chronologic

and physiologic age affect replicative lifespan of fibroblasts from diabetic, prediabetic and normal donors. Science *199:* 781–783 (1978).
Goldstein, S.; Moerman, E.J.; Porter K.: High voltage electronmicroscopy of human diploid fibroblasts during aging in vitro. Expl Cell Res. *154:* 101–111 (1984).
Goldstein, S.; Singal, D.P.: Alteration of fibroblast gene products in vitro from a subject with Werner's syndrome. Nature, Lond. *251:* 719–721 (1974).
Goldstein, S.; Stolland, D.; Cordeiro, R.A.J.: Decreased proteolysis and increased amino acid efflux in aging human fibroblasts. Mech. Age. Dev. *5:* 221–233 (1976).
Goradia, R.Y.; Davies, B.K.: Asymmetry in sister chromatids of human chromosomes. J. Hered. *69:* 369–372 (1978).
Gordon, P.B.; Conn, G.; Hatcher, V.B.: Glycosaminoglycan production in cultures of early and late passage human endothelial cells: the influence of an anionic endothelial cell growth factor and the extracellular matrix. J. cell. Physiol. *125:* 596–607 (1985).
Gorman, S.D.; Cristofalo, V.J.: Analysis of the G1 arrest position of senescent WI-38 cells by quinacrine dihydrochloride nuclear fluorescence. Evidence for a late G1 arrest. Expl Cell Res. *167:* 87–94 (1986).
Gospodarowicz, D.; Bialecki, H.: The effects of the epidermal and fibroblast growth factors on the replicative lifespan of cultured bovine granulosa cells. Endocrinology *103:* 854–865 (1978).
Gospodarowicz, D.; Hirabayashi, K.; Giguere, L.; Tauber, J.P.: Factors controlling the proliferative rate, final cell density and life span of bovine vascular smooth muscle cells in culture. J. Cell Biol. *89:* 568–578 (1981).
Green, H.: Terminal differentiation of cultured human epidermal cells. Cell *11:* 405–416 (1977).
Green, L.; Whittle, W.; Dell'Orco, R.T.; Stein, G.; Stein, J.: Histone gene stability during cellular senescence. Mech. Age. Dev. *36:* 211–215 (1986).
Greenberg, S.B.; Grove, G.L.; Cristofalo, V.J.: Cell size in aging monolayer cultures. In Vitro *13:* 297–300 (1977).
Griffin, J.E.; Allman, D.R.; Durrant, J.L.; Wilson, J.D.: Variation in steroid 5 alpha-reductase activity in cloned human skin fibroblasts. Shift in phenotypic expression from high to low activity upon subcloning. J. biol. Chem. *256:* 3662–3666 (1981).
Griffiths, T.D.: S-phase transit times as a function of age in human diploid fibroblasts. Mech. Age. Dev. *24:* 273–282 (1984).
Griffiths, T.D.; Carpenter, J.; Ling, S.Y.: DNA chain growth as a function of age in intact and permeabilized WI-38 and MRC-5 cells. Mech. Age. Dev. *21:* 15–25 (1983).
Grove, G.L.; Cristofalo, V.J.: The 'transition probability model' and the regulation of proliferation of human diploid cell cultures during aging. Cell Tiss. Kinet. *9:* 395–399 (1976).
Grove, G.L.; Kligman, A.M.: Age-associated changes in human epidermal cell renewal. J. Geront. *38:* 137–142 (1983).
Guernsey, D.L.; Koebbe, M.; Thomas, J.E.; Myerly, T.K.; Zmolek, D.: An altered response in the induction of cell membrane (Na and K) ATPase by thyroid hormone is characteristic of senescence in cultured human fibroblasts. Mech. Age. Dev. *33:* 283–293 (1986).
Gupta, R.S.: Senescence of cultured human diploid fibroblasts. Are mutations responsible? J. cell. Physiol. *103:* 209–216 (1980).

Hahn, G.M.; King, D.; Yang, S.J.: Quantitative changes in unscheduled DNA synthesis in rat muscle cells after differentiation. Nature New Biol. *230:* 242–244 (1971).

Hakami, N.; Pious, D.A.: Mitochondrial enzyme activity in 'senescent' and virus transformed human fibroblasts. Expl Cell Res. *53:* 135–138 (1968).

Hall, J.D.; Almy, R.E.; Scherer, K.L.: DNA repair in cultured human fibroblasts does not decline with donor age. Expl Cell Res. *139:* 351–359 (1982).

Hall, K.Y.; Hart, R.W.; Benirschke, A.K.; Walford, R.: Correlation between ultraviolet-induced DNA repair in primate lymphocytes and fibroblasts and species maximum achievable life span. Mech. Age. Dev. *24:* 163–173 (1984).

Hamelin, C.; Cousineau, L.; Dion, M.; Yelle, J.: Increased DNA topoisomerase-1 activity in aging human cell chromatin. Biosci. Rep. *4:* 861–868 (1984).

Hanaoka, F.; Takeuchi, F.; Matsumura, T.; Goto, M.; Miyamoto, M.; Yamada, H.: Decrease in the average size of replicons in a Werner's syndrome cell line by simian virus 40 infection. Expl Cell Res. *144:* 363–467 (1983).

Hards, R.G.; Patterson, D.: Variation of glycinamide ribonucleotide synthetase levels during in vitro aging of human fibroblasts. Implications for gene dosage studies. Mech. Age. Dev. *36:* 65–70 (1986).

Harley, C.R.; Goldstein, S.: Cultured human fibroblasts. Distribution of cell generations and a critical limit. J. cell. Physiol. *97:* 509–516 (1978).

Harley, C.B.; Pollard, J.W.; Chamberlain, J.W.; Stanners, C.P.; Goldstein, S.: Protein synthetic errors do not increase during aging of cultured human fibroblasts. Proc. natn. Acad. Sci. USA *77:* 1885–1889 (1980).

Harman, D.: Aging. A theory based on free radical and radiation chemistry. J. Geront. *11:* 298–300 (1956).

Harnden, D.G.; Benn, P.A.; Oxford, J.M.; Taylor, A.M.R.; Webb, T.P.: Cytogenetically marked clones in human fibroblasts cultured from normal subjects. Som. Cell Genet. *2:* 55–62 (1976).

Harper, R.A.; Grove, G.L.: Human skin fibroblasts derived from papillary and reticular dermis: differences in growth potential in vitro. Science *204:* 526–527 (1979).

Harris, H.; Sidebottom, E.; Grace, D.M.; Bramwell: The expression of genetic information. A study with hybrid animal cells. J. Cell Sci. *4:* 499–508 (1969).

Harris, H.; Watkins, J.F.; Ford, C.E.; Schoefl, G.I.: Artificial heterokaryons of animal cells from different species. J. Cell Sci. *1:* 1–9 (1966).

Harris, H.: Quantitative growth studies with chick myoblasts in glass-substrate cultures. Growth *21:* 149–166 (1957).

Hart, R.W.; Setlow, R.B.: Correlation between deoxyribonucleic acid excision-repair and life span in a number of mammalian species. Proc. natn. Acad. Sci. USA *71:* 2169–2173 (1974).

Hart, R.W.; Setlow, R.B.: DNA repair in late passage human cells. Mech. Age. Dev. *5:* 67–77 (1976).

Hart, R.W.; Sacher, G.A.; Hoskins, T.L.: DNA repair in a short and a long lived rodent species. J. Geront. *34:* 808–817 (1979).

Harrison, D.E.; Astle, C.M.; Delaittre, J.A.: Loss of proliferative capacity in immunohemopoietic stem cells caused by serial transplantation rather than aging. J. exp. Med. *147:* 1526–1531 (1978).

Hasegawa, N.; Hanaoka, F.; Hori, T.; Yamada, M.: Reevaluation of DNA chain elongation rate in human diploid fibroblasts. Expl Cell Res. *140:* 443–446 (1982).

Hasegawa, N.; Hanaoka, F.; Yamada, M.: Does capacity of DNA replication change during in vitro ageing. Expl Cell Res. *156:* 478–486 (1985).

Haslam, R.J.; Goldstein, S.: Adenosine 3′:5′-cyclic monophosphate in young and senescent human fibroblasts during growth and stationary phase in vitro. Biochem. J. *144:* 253–263 (1974).

Hayflick, L.: The limited in vitro lifetime of human diploid cell strains. Expl Cell Res. *37:* 614–636 (1965).

Hayflick, L.: Oncogenesis in vitro. J. natn. Canc. Inst. Monogr. *26:* 355–385 (1967).

Hayflick, L.; Moorhead, P.S.: The serial cultivation of human diploid cell strains. Expl Cell Res. *25:* 585–621 (1961).

Hedman, K.; Johansson, S.; Vartio, S.; Kjellen, L.; Vaheri, A.; Höök, M.: Structure of the pericellular matrix: association of heparan and chondroitin sulfates with fibronectin-procollagen fibers. Cell *28:* 663–671 (1982).

Hennis, H.L.; Braid, H.L.; Vincent, R.A., Jr.: Unscheduled DNA synthesis in cells of different shape in fibroblast cultures from donors of various ages. Mech. Age. Dev. *16:* 355–361 (1981).

Hildebran, J.N.; Absher, M.; Low, R.B.: Altered rates of collagen synthesis in in vitro aged human lung fibroblasts. In Vitro *19:* 307–314 (1983).

Hill, B.T.: Influence of age on chromatin transcription in murine tissues using an heterologous and an homologous RNA polymerase. Gerontology *22:* 111–123 (1976).

Hill, B.T.; Whelan, R.D.H.; Whatley, S.: Evidence that transcription changes in ageing cultures are terminal events occurring after the expression of a reduced replicative potential. Mech. Age. Dev. *8:* 85–95 (1978).

Hill, D.J.; Strain, A.J.; Elstow, S.F.; Swenne, I.; Milner, R.D.G.: Bi-functional action of transforming growth factor-beta on DNA synthesis in early passage human fetal fibroblasts. J. cell. Physiol. *128:* 322–328 (1986).

Hiu, I.J.; Liepkalns, B.; Azzarone, B.; Macieira-Coelho, A.: Metabolic changes in fibroblast adhesion sites induced by growth factors. Cell Biol. int. Rep. *9:* 379–387 (1985).

Hoehn, E.M.; Bryant, E.M.; Martin, G.M.: The replicative lifespans of euploid hybrids derived from short-lived and long-lived human skin fibroblast cultures. Cytogenet. Cell Genet. *21:* 282–295 (1978).

Hoehn, H.; Simpsom, M.; Bryant, E.M.; Rabinovitch, P.S.; Salk, D.; Martin, G.M.: Effect of chromosome constitution on growth and longevity of human skin fibroblast cultures. Am. J. med. Genet. *7:* 141–154 (1980).

Holland, J.J.; Kohne, D.; Doyle, M.V.: Analysis of virus replication in ageing human fibroblast cultures. Nature, Lond. *245:* 316–318 (1973).

Hollenberg, M.D.; Schneider, E.L.: Receptors for insulin and epidermal growth factor-urogastrone in adult human fibroblasts do not change with donor age. Mech. Age. Dev. *11:* 37–43 (1979).

Holliday, R.: Strong effects of 5-azacytidine on the in vitro lifespan of human diploid fibroblasts. Expl Cell Res. *166:* 543–552 (1986).

Holliday, R.; Huschtscha, L.I.; Tarrant, G.M.; Kirkwood, T.B.L.: Testing the commitment theory of cellular aging. Science *198:* 366–372 (1977).

Holliday, R.; Porterfield, J.S.; Gibbs, D.D.: Premature ageing and occurance of altered enzyme in Werner's syndrome fibroblasts. Nature, Lond. *248:* 762–763 (1974).

Holliday, R.; Tarrant, G.M.: Altered enzymes in ageing human fibroblasts. Nature, Lond. *238:* 26–30 (1972).

Holm-Pedersen, P.; Fenstad, A.M.; Folke, E.A.: DNA, RNA and protein synthesis in healing wounds in young and old mice. Mech. Age. Dev. *3:* 173–185 (1974).

Holt, P.R.; Kotler, D.P.; Yeh, K.: Gut adaptation to nutritional intake in the aging rat. 1. Changes in intestinal structure and cell generation (submitted for publ.).

Holt, P.R.; Pascal, R.R.; Kotler, D.P.: Effect of aging upon small intestinal structure in the Fischer rat. J. Geront. *39:* 642–647 (1984).

Honda, S.; Matsuo, M.: The sensitivity to hyperbaric oxygen of human diploid fibroblasts during ageing in vitro. Mech. Age. Dev. *12:* 31–37 (1980).

Hopkins, N.K.; Sun, F.F.; Gorman, R.R.: Thromboxane A2 biosynthesis in human lung fibroblasts WI-38. Biochem. Biophys. Res. Commun. *85:* 827–836 (1978).

Hori, Y.; Perkins, E.H.; Halsall, M.K.: Decline in phytohemagglutinin responsiveness of spleen cells from aging mice. Proc. Soc. exp. Biol. Med. *144:* 48–53 (1973).

Hornberger, E.; Plotkin, S.A.: Human skin fibroblasts are nonpermissive to coxsakie B4 infection: an age-dependent phenomenon. Intervirology *19:* 195–200 (1983).

Hornsby, P.J.: Regulation of cytochrome P-450 supported 11 beta-hydroxylation of deoxycortisol by steroids, oxygen and antiooxidants in adrenocortical cell cultures. J. biol. Chem. *255:* 4020–4027 (1980).

Hornsby, P.J.: The role of vitamin E in cellular energy metabolism in cultured adrenocortical cells. J. cell. Physiol. *216:* 112–207 (1982).

Hornsby, P.J.; Aldern, K.A.; Harris, S.E.: Adrenocortical cultures as model systems for investigating cellular aging; in Armstrong, Sohal, Cutler, Slater, Free radicals in molecular biology, aging and disease, pp. 203–222 (Raven Press, New York 1984).

Hornsby, P.J.; Gill, G.N.: Regulation of glutamine and pyruvate oxidation in cultured adrenocortical cells by cortisol, antioxidants and oxygen. Effects on cell proliferation. J. cell. physiol. *109:* 111–120 (1981).

Hornsby, P.J.; Harris, S.E.: Oxidative damage to DNA and replicative lifespan in cultured adrenocortical cells. Expl Cell Res. *168:* 203–217 (1987).

Hornsby, P.J.; Simonian, M.H.; Gill, G.N.: Aging of adrenocortical cells in culture. Int. Rev. Cytol., suppl. 10, pp. 131–162 (1979).

Houben, A.; Remacle, J.: Lysosomal and mitochondrial heat labile enzymes in ageing human fibroblasts. Nature, Lond. *275:* 59–60 (1978).

Houben, A.; Houbion, A.; Remacle, J.: Lysosomal and mitochondrial heat labile enzymes in Werner's syndrome fibroblasts. Exp. Gerontol. *15:* 629–631 (1980).

Houben, A.; Raes, M.; Houbion, A.; Remacle, J.: Alteration of enzymes in aging human fibroblasts. I. Conditions for the appearance of an alteration in glucose-6-phosphate dehydrogenase. Mech. Age. Dev. *25:* 35–45 (1984a).

Houben, A.; Raes, M.; Houbion, A.; Remacle, J.: Alteration of enzymes in aging human fibroblasts in culture. II. Conditions for the reversibility and the mechanism of the alteration of glucose-6-phosphate dehydrogenase. Mech. Age. Dev. *25:* 35–45 (1984b).

Houck, J.C.; Sharma, V.K.; Hayflick, L.: Functional failures of cultured human diploid fibroblasts after continued population doublings. Proc. Soc. exp. Biol. Med. *137:* 331–333 (1971).

Houde, M.: H1 histone expression during replicative aging of human diploid fibroblasts; diss. University of Ontario, Hamilton, Canada (1986).

Hunter, S.A.; Burnstein, S.; Cedor, C.: Stimulation of prostaglandin synthesis in WI-38 human lung fibroblasts following inhibition of phospholipid acylation by *p*-hydroxymercuribenzoate. Biochim. biophys. Acta *793:* 202–212 (1984).

Icard, C.; Macieira-Coelho, A.: Resistance to caffeine of mouse fibroblasts after acquisition of an infinite division potential. Cell Biol. int. Rep. *5:* 9–13 (1981).

Icard, C.; Beaupain, R.; Diatloff, C.; Macieira-Coelho, A.: Effect of low dose rate irradiation on the division potential of cells in vitro. VI. Changes in DNA and in radiosensitivity during aging of human fibroblasts. Mech. Age. Dev. *11:* 269–278 (1979).

Icard-Liepkalns, C.; Doly, J.; Macieira-Coelho, A.: Gene reorganization during serial divisions of normal human cells. Biochem. biophys. Res. Commun. *141:* 112–123 (1986).

Icard-Liepkalns, C.; Macieira-Coelho, A.: Aging and hydrocortisone effects on transient structures of replicative DNA of human fibroblasts. Proc. Soc. exp. Biol. Med. *170:* 373–377 (1982).

Ide, T.; Tsuji, Y.; Ishibashi, S.; Mitsui, Y.: Reinitiation of host DNA synthesis in senescent human diploid cells by infection with SV40. Expl Cell Res. *143:* 343–349 (1983).

Ide, T.; Tsuji, Y.; Mitsui, Y.; Ishibashi, S.: Induction of host DNA synthesis in senescent human diploid fibroblasts by infection with human cytomegalovirus. Mech. Age. Dev. *25:* 227–235 (1984a).

Ide, T.; Tsuji, Y.; Nakashima, T.; Ishibashi, S.: Progress of aging in human diploid cells transformed with a ts A mutant of simian virus 40. Expl Cell Res. *150:* 321–328 (1984b).

Inoue, T.; Cronkite, E.P.: The influence of in vivo incubation of aged murine spleen colony-forming units on their proliferative capacity. Mech. Age. Dev. *23:* 177–190 (1983).

Jackson, D.A.; Cook, P.R.: Active genes and their transcripts are closely associated with the nuclear cage. Biol. Cell. *45:* 334 (1982).

Jacobson-Kram, D.; Roe, J.L.; Williams, J.R.; Gange, R.W.; Parrish, J.A.: Decreased in vitro lifespan of fibroblasts derived from skin exposed to photochemotherapy in vivo. Lancet *ii:* 1399–1400 (1982).

Johnson, J.E., Jr.: Fine structure of IMR-90 cells in culture as examined by scanning and transmission electron microscopy. Mech. Age. Dev. *10:* 405–443 (1979a).

Johnson, B.D.; Page, R.C.; Narayanan, A.S.; Pieters, H.P.: Effects of donor age on protein and collagen synthesis in vitro by human diploid fibroblasts. Lab. Invest. *55:* 490–496 (1986).

Johnson, L.F.; Abelson, H.T.; Penman, S.; Green, H.: The relative amounts of the cytoplasmic RNA species in normal, transformed and senescent cultured cell lines. J. cell. Physiol. *90:* 465–470 (1977).

Johnson, L.K.; Longenecker, J.P.: Senescence of aortic endothelial cells in vitro: influence of culture conditions and preliminary characterization of the senescent phenotype. Mech. Age. Dev. *18:* 1–18 (1982).

Johnson, L.V.; Walsh, M.L.; Chen, L.B.: Localization of mitochondria in living cells with rhodamine 123. Proc. natn. Acad. Sci. USA *77:* 990–994 (1980).

Johnson, R.; Strehler, B.L.: Loss of genes coding for ribosomal RNA in ageing brain cells. Nature. Lond. *240:* 412–414 (1972).

Jongkind, J.F.; Verkerk, A.; Visser, W.J.; Dongen, J.M. van: Isolation of autofluorescent 'aged' human fibroblasts by flow sorting. Morphology, enzyme activity and proliferative capacity. Expl Cell Res. *138:* 409–417 (1982).

Kaftory, A.; Weissman-Shomer, P.; Fry, M.: DNA binding proteins in growing and resting young and senescent cultured chick fibroblasts. Mech. Age. Dev. *8:* 75–84 (1978).

Kahn, A.; Guillouzo, A.; Leibovitch, M.P.; Cottreau, D.; Bourel, M.; Dreyfus, J.C.: Heat lability of glucose-6-phosphate dehydrogenase in some senescent human cultured cells. Evidence for its post-synthetic nature. Biochem. biophys. Res. Commun. *77:* 760–766 (1977).

Kahn, A.; Meienhofer, M.C.; Guillouzo, A.; Cottreau, D.; Baffet, G.; Henry, J.; Dreyfus, J.C.: Modifications of phosphoproteins and protein kinases occurring with in vitro aging of cultured human cells. Gerontology *28:* 360–370 (1982).

Kaji, K.; Matsuo, M.: Aging of chick embryo fibroblasts in vitro. I. Saturation density and population doubling rate. Exp. Gerontol. *13:* 439–445 (1978).

Kaji, K.; Matsuo, M.: Aging of chick embryo fibroblasts in vitro. V. Time course studies on polyploid nucleous accumulation. Expl Cell Res. *131:* 410–412 (1981).

Kaji, K.; Matsuo, M.: Responsiveness of human lung diploid fibroblast ageing in vitro to epidermal growth factor: saturation density and lifespan. Mech. Age. Dev. *22:* 129–133 (1983).

Kang, A.H.: Fibroblast activation. J. Lab. clin. Med. *92:* 1–4 (1978).

Kantor, G.J.; Mulkie, J.R.; Hull, D.R.: A study of the effect of ultraviolet light on the division potential of human diploid fibroblasts. Expl Cell Res. *113:* 283–294 (1978).

Kanungo, M.S.: Biochemistry of ageing (Academic Press, New York 1980).

Kapp, L.N.; Klevecz, R.R.: The cell cycle of low passage and high passage human diploid fibroblasts. Expl Cell Res. *101:* 154–158 (1976).

Karatza, C.; Shall, S.: The reproductive potential of normal mouse embryo fibroblasts during culture in vitro. J. Cell Sci. *66:* 401–410 (1984).

Kato, H.; Harada, M.; Tsuchiya, K.; Moriwaki, K.: Absence of correlation between DNA repair in ultraviolet irradiated mammalian cells and life span of the donor species. Jap. J. Genet. *55:* 99–108 (1980).

Kator, K.; Cristofalo, V.J.; Charpentier, R.; Cutler, R.G.: Dysdifferentiative nature of aging: passage number dependency of globin gene expression in normal human diploid cells grown in tissue culture. Gerontology *31:* 355–361 (1985).

Kaufman, M.; Pinsky, L.; Straisfeld, C.; Shanfield, B.; Zilahi, B.: Qualitative differences as an indication of cellular heterogeneity in fibroblast monolayers derived from human preputial skin. Expl Cell Res. *96:* 31–36 (1975).

Kelley, R.O.; Mann, P.L.; Perdue, B.D.; Marek, L.F.: Reduction of filamin in late passage human diploid fibroblasts (IMR-90). Mech. Age. Dev. *30:* 78–89 (1985).

Kelley, R.O.; Perdue, B.D.: Development of the aging cell surface: structural pattern of gap junction assembly between metabolic mutants and progressively subcultivated human diploid fibroblasts (IMR-90). Expl Geront. *15:* 407–422 (1980).

Kelley, R.O.; Trotter, J.A.; Marek, L.F.; Perdue, B.D.; Taylor, C.B.: Variation in cytoskeletal assembly during spreading of progressively subcultivated human fibroblasts (IMR-90). Mech. Age. Dev. *13:* 127–141 (1980).

Kelley, R.O.; Vogel, K.G.; Crissman, H.A.; Lujan, C.J.; Skipper, B.E.: Development of the aging cell surface. Reduction of gap junction-mediated metabolic cooperation with progressive subcultivation of human embryo fibroblasts (IMR-90). Expl Cell Res. *119:* 127–143 (1979).

Kent, W.M.; Funderburg, F.M.; Culp, L.A.: Proteoglycans in the substractum adhesion sites of human papillary or reticular dermal fibroblasts. Aging in vivo or in vitro. Mech. Age. Dev. *33:* 115–137 (1986).

Keuren, M.L. van; Merril, C.R.; Goldman, D.: Protein variations associated with in vitro aging of human fibroblasts and quantitative limits on the error catastrophe hypothesis. J. Geront. *38:* 645–652 (1983).

Kihara, F.; Tsuji, Y.; Miura, M.; Ishibashi, S.; Ide, T.: Events blocked in prereplicative phase in senescent human diploid cells TIG-1, following serum stimulation. Mech. Age. Dev. *37:* 103–117 (1986a).

Kihara, F.; Ninomiya-Tsuji, J.; Ishibashi, S.; Ide, T.: Failure in S6 protein phosphorylation by serum stimulation of senescent human diploid fibroblasts TIG-1. Mech. Age. Dev. *37:* 27–40 (1986b).

Kondo, H.; Kasuga, H.; Noumura, T.: The heterogeneity of human fibroblasts as determined from the effects of hydrocortisone on cell growth and specific dexamethasone binding. Expl Cell Res. *158:* 342–348 (1985).

Konterman, K.; Bayreuther, K.: The cellular aging of rat fibroblasts in vitro is a differentiation process. Gerontology *25:* 261–274 (1979).

Kopelovitch, L.: Are all normal diploid human cell strains alike? Relevance to carcinogenic mechanisms in vitro. Expl Cell Biol. *50:* 266–270 (1982).

Kratochwill, K.: Organ specificity in mesenchymal induction demonstrated in the embryonic development of the mammary gland of the mouse. Devl Biol. *20:* 46–71 (1969).

Krauss, S.W.; Linn, S.: Changes in DNA polymerase alpha, beta and gamma during the replicative lifespan of cultured human fibroblasts. Biochemistry *21:* 1002–1008 (1982).

Kubbies, M.; Schindler, D.; Hoehn, H.; Rabinovitch, P.S.: BrdU-Hoechst flow cytometry reveals regulation of human lymphocyte growth by donor-age-related growth fraction and transition rate. J. cell. Physiol. *125:* 229–234 (1985).

Kunisada, T.; Yamagishi, H.; Ogita, Z.-I.; Kirakawa, T.; Mitsui, Y.: Appearance of extrachromosomal circular DNAs during in vivo and in vitro ageing of mammalian cells. Mech. Age. Dev. *29:* 89–99 (1985).

Kurobe, M.; Furukawa, S.; Hayashi, K.: Synthesis and secretion of an epidermal growth factor by human fibroblast cells in culture. Biochem. Biophys. Res. Commun. *131:* 1080–1085 (1985).

La Belle, M.; Linn, S.: DNA repair in cultured mouse cells of increasing population doubling level. Mutation Res. *132:* 51–61 (1984).

Lamberigts, G.; Dierickx, P.; Moor, P. de; Verhoeven, G.: Comparison of the metabolism and receptor binding of testosterone and 17-beta-hydroxy alpha-and-rostan-3-one in normal skin fibroblast cultures. Influence of origin and passage number. J. clin. Endocr. Metab. *48:* 924-930 (1979).

Land, H.; Parada, L.F.; Weinberg, R.A.: Tumorigenic conversion of primary embryo fibroblasts requires at least two cooperating oncogenes. Nature, Lond. *304:* 596–602 (1983).

La Rocca, P.; Rafferty, K.A., Jr.: Kinetics of chick embryo cell types in culture. J. cell. Physiol. *113:* 203–210 (1982).

Le Borgne de Kaouel, C.; Billard, C.; Macieira-Coelho, A.: Growth characteristics in vitro of hybrids between normal and transformed cell lines. Int. J. Cancer *21:* 338–347 (1978).

Le Guilly, Y.; Simon, M.; Lenoir, P.; Bourel, M.: Long-term culture of human adult liver cells: morphological changes related to in vitro senescence and effect of donor's age on growth potential. Gerontologia *19:* 303–313 (1973).

Leberghe, N.V.; Gansen, P., van: Behaviour and ultrastructure of in vitro ageing mouse embryo fibroblasts grown in collagen. Mech. Age. Dev. *34:* 133–150 (1986).

Lee, S.C.; Bemiller, P.M.; Bemiller, J.N.; Pappelis, A.J.: Nuclear area changes in senescing human diploid fibroblasts. Mech. Age. Dev. *7:* 417–424 (1978).

Lee, H.C.; Paz, M.A.; Gallop, P.M.: Low density lipoprotein receptor binding in aging human diploid fibroblasts in culture. J. biol. Chem. *257:* 8912–8927 (1982).

Lemonnier, F.; Gautier, M.; Dinh, F.N.: Serum albumin synthesis by human fibroblasts in culture. Biochimie *61:* 483–486 (1979).

Lemonnier, F.; Gautier, M.; Wolfrom, C.; Lemonnier, A.: Some metabolic differences between human skin and aponeurosis fibroblasts in culture. J. cell. Physiol. *104:* 415–423 (1980).

Lesher, S.; Fry, R.J.M.; Kohn, H.I.: Aging and the generation cycle of intestinal epithelial cells in the mouse. Gerontologia *5:* 176–181 (1961).

Lewis, C.M.; Tarrant, G.M.: Error theory and aging in human diploid fibroblasts. Nature, Lond. *239:* 316–318 (1972).

Lima, L.; Macieira-Coelho, A.: Parameters of aging in chicken embryo fibroblasts cultivated in vitro. Expl Cell Res. *70:* 279–284 (1972).

Lincoln, D.W.; Karen, I.; Braunschweiger, K.; Braunschweiger, W.R.; Smith, J.R.: The two-dimensional polypeptide profile of terminally non-dividing human diploid cells. Expl Cell Res. *154:* 136–146 (1984).

Lindgren, U.; Farber, J.R.A.: Chromosome replication in aging human diploid fibroblasts. Expl Cell Res. *142:* 301–308 (1982).

Linn, S.; Kairis, M.; Holliday, R.: Decreased fidelity of DNA polymerase activity isolated from aging human fibroblasts. Proc. natn. Acad. Sci. USA *73:* 2818–2822 (1976).

Lipetz, J.; Cristofalo, V.J.: Ultrastructural changes accompanying the aging of human diploid cells in culture. J. ultrastruct. Res. *39:* 43–56 (1972).

Lipman, R.D.; Muggleton-Harris, A.: Modification of the cataractous phenotype by somatic cell hybridization. Som. Cell Genet. *8:* 791–800 (1982).

Lipschitz, D.A.; Udupa, K.B.: Effect of donor age on long-term culture of bone marrow in vitro. Mech. Age. Dev. *24:* 119–127 (1984).

Little, J.B.: Relationship between DNA repair capacity and cellular aging. Gerontology *22:* 28–39 (1976).

Litwin, J.: The effect of light on the aging of human diploid fibroblasts. Exp. Gerontol. *7:* 381–386 (1972).

Liu, A.Y.C.; Chang, Z.F.; Chen, K.Y.: Increased level of cAMP-dependent protein kinase in aging human lung fibroblasts. J. cell. Physiol. *128:* 149–154 (1986).

Liu, S.C.C.; Parsons, C.S.; Hanawalt, P.: DNA repair response in human epidermal keratinocytes from donors of different age. J. invest. Derm. *79:* 330–335 (1982).

Lu, Z.; Dice, J.F.: Ginseng extract inhibits protein degradation and stimulates protein synthesis in human fibroblasts. Biochem. biophys. Res. Commun. *126:* 636–640 (1985).

Lumpkin, C.K., Jr.; Mcllung, J.K.; Pereira-Smith, O.M.; Smith, J.R.: Existence of high abundance antiproliferative mRNAs in senescent human diploid fibroblasts. Science *232:* 393–394 (1986a).

Lumpkin, C.K.; Knepper, J.E.; Butel, J.S.; Smith, J.R.; Pereira-Smith, O.M.: Mitogenic effects of the proto-oncogene and oncogene forms of c-H-ras DNA in human diploid fibroblasts. Mol. cell. Biol. *6:* 2990–2993 (1986b).

Macieira-Coelho, A.: Action of cortisone on human fibroblasts in vitro. Experientia *22:* 390–391 (1966).

Macieira-Coelho, A.: Influence of cell density on growth inhibition of human fibroblasts in vitro. Proc. Soc. exp. Biol. Med. *125:* 548 (1967a).

Macieira-Coelho, A.: Relationship between DNA synthesis and cell density in normal and virus-transformed cells. Int. J. Cancer *2:* 297–303 (1967b).

Macieira-Coelho, A.: Dissociation between inhibition of movement and inhibition of division in RSV transformed human fibroblasts. Expl Cell Res. *47:* 193–200 (1967c).

Macieira-Coelho, A.: The decreased growth potential in vitro of human fibroblasts of adult origin; in Holeckova, Cristofalo, Aging in cell and tissue culture, pp. 121–132 (Plenum Press, New York 1970).

Macieira-Coelho, A.: Aging and cell division. Front. Matrix Biol., vol. I, pp. 46–77 (Karger, Basel 1973a).

Macieira-Coelho, A.: Cell cycle analysis in mammalian cells; in Kruse, Patterson, Tissue culture: methods and application, pp. 412–417 (Academic Press, New York 1973b).

Macieira-Coelho, A.: Are non-dividing cells present in ageing cell cultures? Nature, Lond. *248:* 421–422 (1974).

Macieira-Coelho, A.: Kinetics of the proliferation of human fibroblasts during their lifespan in vitro. Mech. Age. Dev. *6:* 341–343 (1977).

Macieira-Coelho, A.: Reorganization of the cell genome as the basis of aging in dividing cells; in Orimo, Shimada, Iriki, Maeda, Recent advances in gerontology. Proc. XI Int. Congr. Geront., pp. 111–112 (Excerpta Medica, Amsterdam 1979).

Macieira-Coelho, A.: Implications of the reorganization of the cell genome for aging or immortalization of dividing cells in vitro. Gerontology *26:* 276–282 (1980).

Macieira-Coelho, A.: Possible implications of the reorganization of the cell genome for the transfer of information in dividing cells; in Danon, Shock, Marois, Aging: a challenge to science and society, vol. I. Biology, pp. 136–145 (Oxford University Press, London 1981).

Macieira-Coelho, A.: Changes in membrane properties associated with cellular aging. Int. Rev. Cytol. *83:* 183–220 (1983).

Macieira-Coelho, A.: Genome reorganization during cellular senescence. Mech. Age. Dev. *27:* 257–262 (1984).

Macieira-Coelho, A.; Avrameas, S.: Modulation of cell behavior in vitro by the substratum in fibroblastic and leukemic mouse cell lines. Proc. natn. Acad. Sci. USA *69:* 2469–2473 (1972).

Macieira-Coelho, A.; Azzarone, B.: Aging of human fibroblasts is a succession of subtle changes in the cell cycle and has a final short stage with abrupt events. Expl Cell Res. *141:* 325–332 (1982).

Macieira-Coelho, A.; Bengtsson, A.; Vander Ploeg, U.: Distribution of DNA between sister cells during serial subcultivation of human fibroblasts. Histochemistry *75:* 11–24 (1982).

Macieira-Coelho, A.; Berumen, L.: The cell cycle during growth inhibition of human embryonic fibroblasts in vitro. Proc. Soc. exp. Biol. Med. *144:* 43–48 (1973).

Macieira-Coelho, A.; Berumen, L.; Avrameas, S.: Properties of protein polymers as substratum for cell growth in vitro. J. cell Physiol. *83:* 379–388 (1974).

Macieira-Coelho, A.; Bournoutian, C.; Avrameas, S.: The role of the cell membrane in the expression of contact inhibition of division. Biol. Cell *32:* 121–126 (1978).

Macieira-Coelho, A.; Diatloff, C.; Malaise, E.: Effect of low dose rate irradiation on the division potential of cells in vitro. II: Mouse lung fibroblasts. Expl Cell Res. *100:* 228–232 (1976a).

Macieira-Coelho, A.; Diatloff, C.; Malaise, J.E.: Doubling potential of fibroblasts from different species after ionising radiation. Nature, Lond. *261:* 586–588 (1976b).

Macieira-Coelho, A.; Diatloff, C.; Billardon, C.; Bourgeois, C.A.; Malaise, E.: Effect of low dose rate ionizing radiation on the division potential of cells in vitro. III. Human lung fibroblasts. Expl Cell Res. *104:* 215–221 (1977).

Macieira-Coelho, A.; Diatloff, C.; Billard, M.; Fertil, B.; Malaise, E.; Fries, D.: Effect of low dose rate irradiation on the division potential of cells in vitro. IV. Embryonic and adult human lung fibroblast-like cells. J. cell. Physiol. *95:* 235–238 (1978).

Macieira-Coelho, A.; Garcia-Giralt, E.; Adrian, M.: Changes in lysosomal associated structures in human fibroblasts kept in resting phase. Proc. Soc. exp. Biol. Med. *138:* 712–718 (1971).

Macieira-Coelho, A.; Icard-Lipkalns, C.; Puvion-Dutilleul, F.: Genome reorganization during senescence of dividing cells; in Modern trends in aging research, Colloque INSERM/EURAGE, vol. 147, pp. 41–54, John Libbey Eurotext Ltd., Paris (1987).

Macieira-Coelho, A.; Lima, L.: Parameters of aging in chicken embryo fibroblasts cultivated in vitro. Expl Cell Res. *70:* 279–284 (1972).

Macieira-Coelho, A.; Lima, L.: Aging in vitro. Incorporation of RNA and protein precursors and acid phosphatase activity during the lifespan of chick embryo fibroblasts. Mech. Age. Dev. *2:* 13–18 (1973).

Macieira-Coelho, A.; Loria, E.: Stimulation of ribosome synthesis during retarded aging of human fibroblasts by hydrocortisone. Nature, Lond. *251:* 67–69 (1974).

Macieira-Coelho, A.; Loria, E.: Changes in RNA synthesis during the life span of human fibroblasts in vitro. Gerontologia *22:* 79–88 (1976).

Macieira-Coelho, A.; Pontén, J.: Analogy in growth between late passage human embryonic and early passage human adult fibroblasts. J. Cell Biol. *43:* 374–377 (1969).

Macieira-Coelho, A.; Pontén, J.; Philipson, L.: The division cycle and RNA synthesis in diploid human cells at different passage levels in vitro. Expl Cell Res. *42:* 673–684 (1966a).

Macieira-Coelho, A.; Pontén, J.; Philipson, L.: Inhibition of the division cycle in confluent cultures of human fibroblasts in vitro. Expl Cell Res. *43:* 20–29 (1966b).

Macieira-Coelho, A.; Puvion-Dutilleul, F.: Genome reorganization during aging of dividing cells; in Salk, Fujiwara, Martin, Werner's syndrome and human aging (Plenum Press, New York, 1985).

Macieira-Coelho, A.; Taboury, F.: A reevaluation of the changes in proliferation in human fibroblasts during ageing in vitro. Cell Tiss. Kinet. *15:* 213–224 (1982).

Maizel, A.; Nicolini, C.; Baserga, R.: Structural alterations of chromatin in phase III WI-38 human diploid fibroblasts. Expl Cell Res. *96:* 351–359 (1975).

Mallucci, L.; Wells, A.V.; Young, R.: Effect of trypsin on cell volume and mass. Nature New Biol. *239:* 53–33 (1972).

Marcum, J.A.; Conway, E.M.; Youssoufian, H.; Rosenberg, R.D.: Anticoagulant active heparin-like molecules from cultured fibroblasts. Expl Cell Res. *166:* 253–258 (1986).

Martin, G.M.; Sprague, C.A.: Parasexual cycle in cultivated human somatic cells. Science *166:* 761–763 (1969).

Martin, G.M.; Sprague, C.A.; Epstein, C.J.: Replicative lifespan of cultivated human cells. Effects of donor's age, tissue and genotype. Lab. Invest. *23:* 86–92 (1970).

Martin, G.M.; Sprague, C.A.; Norwood, T.H.; Pendergrass, W.R.: Clonal selection, attenuation and differentiation in an in vitro model of hyperplasia. Am. J. Path. *74:* 137–154 (1974).

Martin, G.M.; Ogburn, G.E.; Wight, T.N.: Comparative rates of decline in the primary cloning efficiencies of smooth muscle cells from the aging thoracic aorta of two murine species of contrasting maximum life span potentials. Am. J. Path. *110:* 236–245 (1983).

Martinez, A.O.; Norwood, T.H.; Prothero, J.W.; Martin, G.M.: Evidence for clonal attenuation of growth potential in Hela cells. In Vitro *14:* 996–1001 (1978).

Martinez, A.O.; Vigil, A.; Vila, J.C.: Flow-cytometric analysis of mitochondria-associated fluorescence in young and old human fibroblasts. Expl Cell Res. *164:* 551–555 (1986).

Maslansky, C.J.; Williams, G.M.: Ultraviolet light-induced DNA repair synthesis in hepatocytes from species of differing longevities. Mech. Age. Dev. *29:* 191–203 (1985).

Massie, H.R.; Baird, M.B.; Samis, H.V.: Prolonged cultivation of primary chick cultures using organic buffers. In Vitro *9:* 441–444 (1974).

Matsumura, T.; Hagashi, M.; Konishi, R.: Immortalization in culture of rat cells: a genetic study. J. natn. Cancer Inst. *74:* 1223–1232 (1985).

Matsumura, T.; Miyashita, S.; Ohno, T.: Conversion of proliferation and production of the colony stimulating factor during serial passage of mouse fibroblasts in culture. Cell Struct. Funct. *4:* 267–274 (1979a).

Matsumura, T.; Pfendt, E.A.; Hayflick, L.: DNA synthesis in the human diploid cell strain WI-38 during in vitro aging. An autoradiography study. J. Gerontol. *34:* 323–328 (1979b).

Matsumura, T.; Pfendt, E.A.; Zerrudo, Z.; Hayflick, L.: Senescent human diploid cells (WI-38). Attempted induction of proliferation by infection with SV40 and by fusion with irradiated continuous cell-lines. Expl Cell Res. *125:* 453–457 (1980).

Matsuo, M.; Kaji, K.; Utakoji, T.; Hosoda, K.: Ploidy of human embryonic fibroblasts during in vitro aging. J. Geront. *37:* 33–37 (1982).

Mattern, M.R.; Cerutti, P.A.: Age-dependent excision repair of damaged thymidine from gamma-irradiated DNA by isolated nuclei from human fibroblasts. Nature, Lond. *254:* 450–452 (1975).

Matuoka, K.; Mitsui, Y.: Involvment of cell surface heparan sulfate in density dependent inhibition of cell proliferation. Cell Struct. Funct. *6:* 23–34 (1981).

Matzner, Y.; Gavison, R.; Schlomai, Z.; Ben-Bassat, H.; Liebergal, M.; Robinson, D.R.; Babior, B.M.: Production of C5a antagonist by synovial and peritoneal tissue fibroblasts. J. cell. Physiol. *129:* 215–220 (1986).

Mayer, P.J.; Bradley, M.O.; Nichols, J.W.W.: No change in DNA damage or repair of single- and double-strand breaks as human diploid fibroblasts age in vitro. Expl Cell Res. *166:* 497–509 (1986).

Mayne, L.V.; Priestley, A.; James, M.R.; Burke, J.F.: Efficient immortalization and morphological transformation of human fibroblasts by transfection with SV40 DNA linked to a dominant marker. Expl Cell Res. *162:* 530–538 (1986).

Mazia, D.: Synthetic activities leading to mitosis. J. cell. comp. Physiol. *62:* 123–140 (1963).

McClain, D.; Edelman, G.N.: Density-dependent stimulation and inhibition of cell growth by agents that disrupt microtubules. Proc. natn. Acad. Sci. USA *77:* 2748–2752 (1980).

McCoy, B.J.; Galdun, J.; Cohen, J.K.: Effects of density and cellular aging on collagen synthesis and growth kinetics in keloid and normal skin fibroblasts. In Vitro *18:* 79–86 (1982).

McKeehan, W.L.; Ham, R.G.: Stimulation of clonal growth of normal fibroblasts with substrata coated with basic polymers. J. Cell. Biol. *71:* 727–731 (1976).

Medvedev, Z.A.; Medvedeva, M.N.; Huschtscha, L.I.: Age changes of the pattern of H1 histone subfractions in rat liver and spleen chromatin. Gerontology *23:* 334–341 (1977).

Meek, R.L.; Ribeiro, T.; Daniel, W.: Patterns of nonscheduled DNA synthesis in mouse embryo cells associated with in vitro aging and with spontaneous transformation to a continuous cell line. Expl Cell Res. *129:* 265–271 (1980).

Mellgren, J.: Effects of the number of cell divisions and of added isologous nuclei acids on aging of normal human fibroblasts in vitro. Path. Eur. *10:* 215–219 (1975).

Merz, G.S., Jr.; Ross, J.D.: Viability of human diploid cells as a function of in vitro age. J. cell. Physiol. *74:* 219–223 (1969).

Merz, G.S., Jr.; Ross, J.D.: Clone size variation in the human diploid cell strain WI-38. J. cell. Physiol. *82:* 75–80 (1973).

Mets, T.; Verdonk, G.: The theory of transition probability and the division pattern of WI-38 cells. Cell Biol. int. Rep. *2:* 561–564 (1978).

Mets, T.; Verdonk, G.: Variations in the stromal cell population of human bone marrow during aging. Mech. Age. Dev. *15:* 41–49 (1981).

Mets, T.; Korteweg, M.; Verdonk, G.: Increased prostaglandin F_2 alpha and E_2 production in late passage WI-38 diploid fibroblasts. Cell Biol. int. Rep. *3:* 691–694 (1979).

Mets, T.; Bekaert, E.; Verdonk, G.: Similarity between in vitro and in vivo cellular aging. Mech. Age. Dev. *22:* 71–78 (1983).

Michl, J.; Svobodova, J.: RNA turnover and the growth potential of human cells in culture. Expl Cell Res. *47:* 616–619 (1967).

Milisauskas, V.; Rose, N.R.: Immunochemical quantitation of enzymes in human diploid cell line WI-38. Expl Cell Res. *81:* 279–284 (1973).

Miller, R.C.; Nichols, W.W.; Pottash, J.; Aronson, M.M.: Cytogenetic comparison of diploid human fibroblast and epitheloid cell lines. Expl Cell Res. *110:* 63–73 (1977).

Millis, A.J.T.; Hoyle, M.; Field, B.: Human fibroblasts conditioned media contains growth promoting activities for low density cells. J. cell. Physiol. *93:* 17–24 (1977).

Milo, G.E., Jr.: Enhancement of senescence in low passage human embryonic lung cells by an agent extracted from phase III cells. Expl Cell Res. *79:* 143–151 (1973).

Milo, G.E., Jr.; Casto, B.C.: Conditions for transformation of human fibroblast cells: an overview. Cancer Lett. *31:* 1–13 (1986).

Milo, G.E., Jr.; Hart, R.W.: Age-related alterations in plasma membrane glycoprotein content and scheduled or nonscheduled DNA synthesis. Archs Biochem. Biophys. *176:* 324–333 (1976).

Minkowitz, S.; Conklin, D.P.; Kaplan, P.: Total LD and isozyme variations during in vitro fibroblast aging with a discussion of its theoretical implication. Mech. Age. Dev. *31:* 163–169 (1985).

Miquel, J.; Fleming, J.E.: A two step hypothesis on the mechanisms of in vitro cell aging: cell differentiation followed by intrinsic mithocondrial mutagenesis. Exp. Gerontol. *19:* 31–36 (1984).

Mitsui, Y.; Sakagami, H.; Murota, S.I.; Yamada, M.A.: Age-related decline in histone H_1 fraction in human diploid fibroblast cultures. Expl Cell Res. *126:* 289–298 (1980).

Mitsui, Y.; Schneider, E.L.: Characterization of fractionated human diploid fibroblast populations. Expl Cell Res. *103:* 23–30 (1976a).

Mitsui, Y.; Schneider, E.L.: Relationship between cell replication and volume in senescent human diploid fibroblasts. Mech. Age. Dev. *5:* 45–56 (1976b).

Mollenhauer, J.; Bayreuther, K.: Donor-age related changes of morphology growth potential and collagen biosynthesis in rat fibroblast sub-population in vitro. Differentiation *32:* 165–172 (1986).

Moore, G.E.; McLimans, W.F.: The lifespan of the cultured normal cell. Concepts derived from studies of human lymphoblasts. J. theor. Biol. *20:* 217–226 (1968).

Moore, C.; Schwartz, A.G.: Inverse correlation between species lifespan and capacity of cultured fibroblasts to convert benzopyrene to water soluble metabolites. Expl Cell. Res. *116:* 359–364 (1978).

Muggleton-Harris, A.L.; Aroian, M.A.: Replicative potential of individual cell hybrids derived from young and old donor human skin fibroblasts. Som. Cell Genet. *8:* 41–50 (1982).

Muggleton-Harris, A.L.; Defuria, R.: Age-dependent metabolic changes in cultured human fibroblasts. In Vitro Cell dev. Biol. *21:* 271–276 (1985).

Muggleton-Harris, A.L.; Desimone, D.W.: Replicative potentials of various fusion products between WI-38 and SV40 transformed WI-38 cells and their components. Som. Cell Genet. *6:* 689–698 (1980).

Muggleton-Harris, A.L.; Hayflick, L.: Cellular aging studied by the reconstruction of replicating cells from nuclei and cytoplasms isolated from normal human diploid cells. Expl Cell Res. *103:* 321–330 (1976).

Muggleton-Harris, A.L.; Reisert, P.S.; Burghoff, R.L.: In vitro characterization of response to stimulus (wounding) with regard to ageing in human skin fibroblasts. Mech. Age. Dev. *19:* 37–43 (1982).

Mukherjee, A.B.; Weinstein, M.E.: Culture media variation as related to in vitro aging of human fibroblasts. I. Effects on population doubling, nuclear volume and nuclear morphology. Mech. Age. Dev. *37:* 55–67 (1986).

Mulherkar, R.; Sharma, A.: Chromosomal and cytochemical changes in human fibroblast cultures. A review. J. Sci. ind. Res. *38:* 322–327 (1979).

Murota, S.; Mitsui, Y.; Kawamura, M.: Effect of in vitro aging on 6-ketoprostaglandin F1 alpha-producing activity in cultured human diploid lung fibroblasts. Biochim. biophys. Acta *574:* 351–355 (1979).

Murray, V.: Properties of DNA polymerase from young and aging fibroblasts. Mech. Age. Dev. *16:* 327–343 (1981).

Murray, V.; Holliday, R.: Increased error frequency of DNA polymerase from senescent human fibroblasts. J. mol. Biol. *146:* 55–76 (1981).

Nagata, K.; Humphrey, M.J.; Olden, K.; Yamada, K.M.: Collagen can modulate cell-interactions with fibronectin. J. Cell Biol. *101:* 386–394 (1985).

Namba, M.; Nishitani, K.; Hyodoh, F.; Fuhushima, F.; Kinoto, T.: Neoplastic transformation of human diploid fibroblasts (KMST-6) by treatment with ^{60}Co gamma rays. Int. J. Cancer *35:* 275–282 (1985).

Nette, E.G.; Sit, H.L.; King, L.D.W.: Reactivation of DNA synthesis in aging diploid human skin fibroblasts by fusion with mouse L karyoplasts cytoplasts and whole L cells. Mech. Age. Dev. *18:* 75–87 (1982).

Nette, E.G.; Pingxi, Y.; Sun, Y.; Andrews, A.D.; King, D.: A correlation between aging and DNA repair in human epidermal cells. Mech. Age. Dev. *24:* 283–292 (1984).

Niedermüller, H.; Hofecker, G.; Skalicky, M.: Changes of DNA repair mechanisms during the aging of the rat. Mech. Age. Dev. *29:* 221–238 (1985).

Niewarowski, S.; Goldstein, S. Interaction of cultured human fibroblasts with fibrin: modification by drugs and aging in vitro. J. Lab. clin. Med. *82:* 605–610 (1973).

Nilsson, J.; Ksiazek, T.; Heldin, C.H.; Thyberg, J.: Demonstration of stimulatory effects of platelet-derived growth factor on cultivated rat arterial smooth muscle cells. Expl Cell Res. *145:* 231–237 (1983).

Nirenborg, M.W.; Matthei, J.W.: The dependence of cell-free protein synthesis in *E. coli* upon naturally occurring or synthetic polynucleotides. Proc. natn. Acad. Sci. USA *47:* 1588–1602 (1961).

Norrby, K.; Bergström, S.; Druvefors, P.: Age dependent mitogenesis in normal connective tissue cells. Virchows Arch. Abt. B Zellpath. *36:* 27–34 (1981).

Norwood, T.H.; Pendergrass, W.R.; Sprague, C.A.; Martin, G.M.: Dominance of the senescent phenotype in heterokaryons between replicative and post-replicative human fibroblasts-like cells. Proc. natn. Acad. Sci. USA *71:* 2231–2235 (1974).

Norwood, T.H.; Pendergrass, W.R.; Martin, G.M.: Reinitiation of DNA synthesis in senescent human fibroblasts upon fusion with cells of unlimited growth potential. J. Cell Biol. *64:* 551–556 (1975).

Norwood, T.H.; Pendergrass, W.R.; Bornstein, P.; Martin, G.M.: DNA synthesis of sublethally injured cells in heterokaryons and its relevance to clonal senescence. Expl Cell Res. *119:* 15–21 (1979).

Nose, K.; Okamoto, H.: Transcriptional activity of nuclei from WI-38 cells at various passages. J. cell. Physiol. *102:* 51–54 (1980).

Ohno, T.: Strict relationship between dialyzed serum concentration and cellular lifespan in vitro. Mech. Age. Dev. *11:* 179–183 (1979).

Ohno, T.; Ohkawa, A.; Sato, C.: Factors controlling the appearance of the immunofluorescent nuclear dots revealed with monoclonal antibody against microtubule-associated protein-1. Expl Cell Res. *158:* 558–562 (1986).

Ohno, T.; Yamadi, G.: Growth stimulation by direct cell-cell contact (contact-stimulation) in normal human lung fibroblasts. Biomed. Pharm. *38:* 337–343 (1984).

Okada, Y.; Kimura, H.; Aoyama, T.: Neurotropin increases in vitro life span of human fibroblasts. Mech. Age. Dev. *35:* 133–143 (1986).

Olashaw, N.E.; Kress, E.D.; Cristofalo, V.J.: Thymidine triphosphate synthesis in senescent WI-38 cells. Expl Cell Res. *149:* 547–553 (1983).

Ono, T.; Cutler, R.G.: Age dependent relaxation of gene expression. Increase of endogenous murine leukemia virus-related and globin-related RNA in brain and liver of mice. Proc. natn. Acad. Sci. USA *75:* 4431–4435 (1978).

Pacifici, M.; Fellini, S.A.; Holtzer, H.; Luca, S. de: Changes in the sulfated proteoglycans synthesized by 'aging' chondrocytes. I. Dispersed cultured chondrocytes and in vivo cartilages. J. biol. Chem. *25:* 1029–1037 (1981).

Painter, R.B.; Clarkson, J.M.; Young, B.R.: Ultraviolet-induced repair replication in aging diploid human cells (WI-38). Radiat. Res. *56:* 560–564 (1973).

Parker, R.C.: The races that constitute the group of common fibroblasts. III. Differences determined by origin of explant and age of donor. J. exp. Med. *58:* 401–414 (1933).

Parshad, R.; Sanford, K.K.: Proliferative response of human diploid fibroblasts to intermittent light exposure. J. cell. Physiol. *92:* 481–486 (1977).

Pashko, L.; Schwartz, A.G.: Inverse correlation between species life span and specific cytochrome P-448 content of cultured fibroblasts. J. Geront. *37:* 38–41 (1982).

Paulsson, Y.; Bywater, M.; Pfeifer-Ohlsson, S.; Ohlsson, R.; Nilsson, S.; Heldin, C.H.; Westermark, B.; Betsholtz, C.: Growth factors induce early prereplicative changes in senescent human fibroblasts. Eur. molec. Biol. Org. J. *5:* 2157–2162 (1986).

Paz, M.A.; Gallop, P.M.: Collagen synthesized and modified by aging fibroblasts in culture. In Vitro *11:* 302–312 (1975).

Paz, M.A.; Torrelio, M.; Gallop, P.M.: X-linked processes in serially passaged aging human diploid cells. J. Geront. *36:* 142–151 (1981).

Pederson, T.: Chromatin structure and the cell cycle. Proc. natn. Acad. Sci. USA *69:* 2224–2228 (1972).

Peeter-Joris, C.; Vaes, G.: Degradation of cartilage proteoglycan and collagen by synovial cells. Lymphocyte factors bacterial products or other inflammatory stimuli. Biochem. biophys. Acta *804:* 474–486 (1984).

Peleg, L.; Raz, E.; Ben-Ishai, R.: Changing capacity for DNA excision repair in mouse embryonic cells in vitro. Expl Cell Res. *104:* 301–307 (1976).

Pendergrass, W.R.; Martin, G.M.; Bornstein, P.: Evidence contrary to the protein error hypothesis for in vitro senescence. J. cell. Physiol. *87:* 3–14 (1976).

Pendergrass, W.R.; Saulewicz, A.C.; Burmer, G.C.; Rabinovitch, P.S.; Norwood, T.H.; Martin, G.M.: Evidence that a critical threshold of DNA polymerase alpha activity may be required for the initiation of DNA synthesis in mammalian cell heterokaryons. J. cell. Physiol. *113:* 141–151 (1982).

Pendergrass, W.R.; Saulevicz, A.S.; Salk, D.; Norwood, T.: Induction of DNA polymerase alpha in senescent cultures of normal and Werner's syndrome cultured skin fibroblasts. J. cell. Physiol. *124:* 331–336 (1985).

Pereira, O.M.; Smith, J.R.; Packer, L.: Photosensitization of human diploid cell cultures by intercellular flavins and protection by antioxydants. Photochem. Photobiol. *24:* 237–242 (1976).

Pereira-Smith, O.M.; Smith, J.R.: Expression of SV40 T antigen in finite life-span hybrids of normal and SV40-transformed fibroblasts. Som. Cell Genet. *7:* 411–421 (1981).

Pereira-Smith, O.M.; Smith, J.R.: Phenotype of low proliferative potential is dominant in hybrids of normal human fibroblasts. Som. Cell Genet. *8:* 731–742 (1982).

Pereira-Smith, O.M.; Smith, J.R.: Evidence for the recessive nature of cellular immortality. Science *221:* 964–966 (1983).

Peters, T.D.; Farber, R.A.; Tarrant, G.M.; Holliday, R.: Altered rate of DNA replication in ageing human fibroblast cultures. Nature, Lond. *251:* 434–436 (1974).

Peterson, C.; Goldman, J.E.: Alterations in calcium content and biochemical processes in cultured skin fibroblasts from aged and Alzheimer donors. Proc. natn. Acad. Sci. USA *83:* 2758–2762 (1986).

Peterson, C.; Ratan, R.R.; Shelanski, M.L.; Goldman, J.E.: Cytosolic free calcium and cell spreading decrease in fibroblasts from aged and Alzheimer donors. Proc. natn. Acad. Sci. USA *83:* 7999–8001 (1986).

Phillips, P.D.; Docherty, P.A.; Mitchell, R.B.: Histochemical assessment of nuclear protein content and enzyme activity in confluent nonarrested and low-serum arrested WI-38 fibroblasts at mid and late population doubling levels. Mech. Age. Dev. *20:* 1–12 (1982).

Phillips, P.D.; Kaji, K.; Cristofalo, V.J.: Progressive loss of the proliferation response of senescent WI-38 cells to platelet derived growth factor, epidermal growth factor, insulin, transferrin and dexamethasone. J. Geront. *39:* 11–17 (1984).

Phillips, P.D.; Kuhnle, E.; Cristofalo, V.C.: I-EGF binding ability is stable throughout the replicative life-span of WI-38 cells. J. cell. Physiol. *114:* 311–316 (1983).

Pitha, J.; Stork, E.; Wimmer, E.: Protein synthesis during aging of human cells in culture. Direction by polio virus. Expl Cell Res. *94:* 310–314 (1975).

Plisko, A.; Gilchrest, B.A.: Growth factor responsiveness of cultured human fibroblasts declines with age. J. Geront. *38:* 513–518 (1983).

Poiley, J.A.; Schuman, R.F.; Pienta, R.J.: Characterization of normal human embryo cells grown to over 100 population doublings. In Vitro *14:* 405–412 (1978).

Polgar, P.; Taylor, L.; Brown, L.: Plasma membrane associated metabolic parameters and the aging of human diploid fibroblasts. Mech. Age. Dev. *7:* 151–160 (1978).

Pollack, A.; Bagwell, C.B.; Irvin, G.L., III: Radiation from tritiated thymidine perturbs the cell cycle progression of stimulated lymphocytes. Science *203:* 1025–1027 (1979).

Pontén, J.: The in vivo growth mechanism of avian Rous sarcoma. Natn. Cancer Inst. Monogr. *17:* 131–145 (1964).

Pontén, J.: Spontaneous and virus induced transformation in cell culture. Virology monographs, No. 8, pp. 4–186 (Springer, New York 1971).

Pontén, J.: The growth capacity of normal and Rous virus transformed chicken fibroblasts in vitro. Int. J. Cancer *6:* 323–332 (1970).

Pontén, J.; Stein, W.D.; Shall, S.: A quantitative analysis of the aging of human glial cells in culture. J. cell. Physiol. *117:* 342–352 (1983).

Pool, T.B.: A comparison of protein synthesis between serum deprived and postmitotic human diploid fibroblasts. Cell Biol. int. Rep. *6:* 658 (1982).

Poot, M.; Visser, W.J.; Verkerk, A.; Jongkind, J.F.: Autofluorescence of human skin

fibroblasts during growth inhibition and in vitro ageing. Gerontology *31:* 158–165 (1985).
Poot, M.; Verkerk, A.; Jongkind, J.F.: Glutathione content of cultured human fibroblasts during in vitro aging. Mech. Age. Dev. *27:* 315–321 (1984).
Poot, M.; Verkerk, A.; Jongkind, J.F.: Accumulation of a high molecular weight glycoprotein during in vitro ageing and contact inhibition of growth. Mech. Age. Dev. *34:* 216–232 (1986).
Praeger, F.C.; Gilchrest, B.A.: Influence of increased extracellular calcium concentration and donor age on density-dependent growth inhibition of human fibroblasts. Proc. Soc. exp. Biol. Med. *182:* 315–321 (1986).
Prentice, D.A.; Tobey, R.A.; Gurley, L.R.: Cell cycle variations in chromatin structure detected by DNase I. Expl Cell Res. *157:* 242–252 (1985).
Press, G.D.; Pitha, J.: Aging changes in uptake of polysaccarides by human diploid cells in culture. Mech. Age. Dev. *3:* 323–328 (1974).
Preumont, A.M.; Capone, B.; Van Gansen, P.: Replicative activity and actinomycin binding in mouse diploid fibroblasts (in vitro ageing). Mech. Age. Dev. *22:* 169–177 (1983).
Price, G.B.; Makinodan, T.: Immunologic deficiencies in senescence. I. Characterization of intrinsic deficiencies. J. Immun. *108:* 413–417 (1972).
Puck, T.T.: Cyclic AMP, the microtubule-microfilament system and cancer. Proc. natn. Acad. Sci. USA *74:* 4491–4495 (1977).
Puvion-Dutilleul, F.; Azzarone, B.; Macieira-Coelho, A.: Comparison between proliferative changes and nuclear events during ageing of human fibroblasts in vitro. Mech. Age. Dev. *20:* 75–92 (1982).
Puvion-Dutilleul, F.; Macieira-Coelho, A.: Ultrastructural organization of nucleoproteins during aging of cultured human embryonic fibroblasts. Expl. Cell Res. *138:* 423–429 (1982).
Puvion-Dutilleul, F.; Macieira-Coelho, A.: Aging dependent nucleolar and chromatin changes in cultivated fibroblasts. Cell Biol. int. Rep. *7:* 61–71 (1983).
Puvion-Dutilleul, F.; Puvion, E.; Icard-Lipkalns, C.; Macieira-Coelho, A.: Chromatin structure, DNA synthesis and transcription through the lifespan of human embryonic lung fibroblasts. Expl Cell Res. *151:* 283–298 (1984).
Rabinovitch, P.S.: Regulation of human fibroblast growth rate by both noncycling cell fraction and transition probability is shown by growth in 5-bromodeoxyuridine followed by Hoechst 33258 flow cytometry. Proc. natn. Acad. Sci. USA *80:* 2951–2955 (1983).
Rabinovitch, P.S.; Norwood, T.H.: Comparative heterokaryon study of cellular senescence and the serum deprived state. Expl Cell Res. *130:* 101–109 (1980).
Raes, M.; Brabander, M. de; Remacle, J.: Polyploid cells in ageing hamster fibroblasts in vitro: possible implication of the centrosome. Eur. J. Cell Biol. *35:* 70–80 (1984).
Raes, M.; Genens, G.; Brabander, M. de; Remacle, J.: Microtubules and microfilaments in ageing hamster embryo fibroblasts in vitro. Exp. Gerontol. *18:* 241–254 (1983).
Raes, M.; Houbion, A.; Remacle, J.: The purification of plasma membranes from WI-38 fibroblasts. Effects of aging on their composition. Biochim. biophys. Acta *641:* 313–324 (1981).
Raes, M.; Remacle, J.: Aging of hamster embryo fibroblasts as the result of both differentiation and stochastic mechanisms. Exp. Gerontol. *18:* 223–240 (1983).

Ramaekers, F.C.S.; Hukkelhoven, M.W.A.C.; Groeneveld, A.; Bloemendal, H.: Changing protein patterns during lens cells aging in vitro. Biochim. biophys. Acta *799:* 221–229 (1984).

Rao, M.V.N.: Cell population heterogeneity in the inductibility of DNA synthesis in human diploid fibroblasts. J. Cell Biol. *89:* 194–197 (1981).

Razin, S.; Pfendt, E.A.; Matsumura, T.; Hayflick, L.: Comparison by autoradiography of macro-molecular biosynthesis in 'young' and 'old' human diploid fibroblast cultures. A brief note. Mech. Age. Dev. *6:* 379–385 (1977).

Reeves, R.: Transcriptionally active chromatin. Biochim. biophys. Acta *782:* 343–393 (1984).

Regan, J.D.; Setlow, R.B.: DNA repair in human progeroid cells. Biochem. biophys. Res. Commun. *59:* 858–864 (1974).

Reimer, D.L.; Singh, S.S.: Cyclophosphamide-induced in vivo sister chromatid exchange in mus musculus. II. Effect of age and genotype on sister chromatid exchange, micronuclei and metaphase index. Mech. Age. Dev. *21:* 59–68 (1983).

Reiss, W.; Kanety, H.; Schlessinger, J.: Five enzymes of the glycolytic pathway serve as substrates for purified epidermal growth-factor-receptor kinase. Biochem. J. *239:* 691–697 (1986).

Remacle, J.A.; Houbion, A.; Houben, A.: Subcellular fractionation of WI-38 fibroblasts comparison between young and old cells. Biochim. biophys. Acta *630:* 57–70 (1980).

Resing, K.A.; Walsh, K.A.; Dale, B.A.: Identification of two intermediates during processing of profilaggrin to filaggrin in neonatal mouse experiments. J. Cell Biol. *99:* 1372–2378 (1984).

Rheinwald, J.G.: The role of terminal differentiation in the finite culture lifetime of the human epidermal keratinocyte. Int. Rev. Cytol., suppl. 10, pp. 25–33 (1979).

Rheinwald, J.G.; Green, J.H.: Epidermal growth factor and the multiplication of cultured human epidermal keratinocytes. Nature, Lond. *265:* 421–424 (1977).

Rhim, J.S.; Huebner, R.J.; Arnstein, P.; Kopelovitch, L.: Chemical transformation of cultured human skin fibroblasts derived from individuals with hereditary adenomatosis of the colon and rectum. Int. J. Cancer *26:* 565–569 (1980).

Riabowol, K.T.; Shmoockler Reis, R.J.; Goldstein, S.: Properties of extrachromosomal covalently closed circular DNA isolated and cloned from aged human fibroblasts. Age *8:* 122–127 (1985).

Rice, R.; Green, H.: Presence in human epidermal cells of a soluble protein precursor of the cross-linked enveloppe. Activation of the cross-linking by calcium ions. Cell *18:* 681–694 (1979).

Rittling, S.R.; Brooks, K.M.; Cristofalo, V.J.; Baserga, R.: Expression of cell cycle-dependent genes in young and senescent WI-38 fibroblasts. Proc. natn. Acad. Sci. USA *83:* 3316–3320 (1986).

Robbins, E.; Levine, E.M.; Eagle, H.: Morphologic changes accompanying senescence of cultured human diploid cells. J. exp. Med. *131:* 1211–1222 (1970).

Robert, R.; Robert, L.: Aging of Connective tissues, Front. Matrix Biol., vol. 1, pp. 1–45 (Karger, Basel 1973).

Röhme, D.: Ageing and the fusion sensitivity potential of human cells in culture. Relation to tissue origin, donor age, and in vitro culture level and condition. Mech. Age. Dev. *16:* 241–253 (1981).

Rosen, E.M.; Mueller, S.N.; Noveral, J.P.; Levine, E.H.: Proliferative characteristics of clonal endothelial cell strains. J. cell. Physiol. *137:* 107–123 (1981).
Rosenbloom, A.L.; Goldstein, S.; Yip, C.C.: Insulin binding to cultured human fibroblasts increases with normal and precocious aging. Science *193:* 412–415 (1976).
Ross, R.: The pathogenesis of atherosclerosis. An update. New Engl. J. Med. *314:* 488–500 (1986).
Ross, R.; Raines, E.W.; Bowen-Pope, D.F.: The biology of platelet-derived growth factor. Cell *46:* 155–169 (1986).
Roth, G.S.; Adelman, R.C.: Age-dependent regulation of mammalian DNA synthesis and cell division in vivo by glucocorticoids. Exp. Gerontol. *9:* 27–31 (1974).
Rothstein, M.: Recent developments in the age-related alteration of enzymes. A review. Mech. Age. Dev. *6:* 241–257 (1977).
Rowe, D.W.; Starman, B.J.; Fujimoto, W.Y.; Williams, R.H.: Abnormalities in proliferation and protein synthesis in skin fibroblast cultures from patients with diabetes mellitus. Diabetes *26:* 284–290 (1977a).
Rowe, D.W.; Starman, B.J.; Fujimoto, W.Y.; Williams, R.H.: Differences in growth response to hydrocortisone and ascorbic acid by human diploid fibroblasts. In Vitro *13:* 824–830 (1977b).
Rush, M.G.; Misra, R.: Extrachromosomal DNA in eucaryotes. Plasmid *14:* 177–191 (1985).
Russell, J.; Witt, W.: Cell size and growth characteristics of cultured fibroblasts isolated from normal and keloid tissue. Plast. reconst. Surg. *57:* 207–212 (1976).
Ryan, J.M.: Effect of different fetal bovine serum concentrations on the replicative lifespan of cultures chick cells. In Vitro *15:* 895–899 (1979a).
Ryan, J.M.: The kinetics of chick cell populations aging in vitro. J. cell. Physiol. *99:* 67–78 (1979b).
Ryan, J.M.; Cristofalo, V.J.: Histone acetylation during aging of human cells in culture. Biochem. biophys. Res. Commun. *48:* 735–742 (1972).
Ryan, J.M.; Nielsen, P.J.; Nelson, L.R.: The effects of division rate-limiting amounts of fetal bovine serum on the proliferation and aging of cultured chick cells. J. cell. Physiol. *110:* 175–181 (1982).
Ryan, J.; Duda, G.; Cristofalo, V.J.: Error accumulation and aging in human diploid cells. J. Geront. *29:* 616–621 (1974).
Ryan, J.M.; Ostrow, D.G.; Breaicefield, X.O.; Gershon, E.S.; Upchurch, L.: A comparison of the proliferative and replicative lifespan kinetics of cell cultures derived from monozygotic twins. In Vitro *17:* 20–27 (1981).
Ryan, J.M.; Pace, D.M.: Alterations in the growth rate and metabolism of chinese hamster cells in vitro. Proc. Soc. exp. Biol. Med. *148:* 320–323 (1975).
Sage, H.: Low molecular weight fibroblasts collagen: structure, secretion and differential expression as a functional of fetal and cellular age. Biochemistry *24:* 7430–7439 (1985).
Sager, R.; Tanaka, K.; Lan, C.C.; Ebina, Y.; Anisowicz, A.: Resistance of human cells to tumorigenesis induced by cloned transforming genes. Proc. natn. Acad. Sci. USA *80:* 7601–7605 (1983).
Sakagami, H.; Mitsui, J.; Murota, S.; Yamada, M.: Two dimensional electrophoretic analy-

sis of nuclear acid proteins in senescent human diploid cells. Cell Struct. Funct. *4:* 215–225 (1979).

Sakagami, H.; Yamada, M.: Failure of vitamin E to extend the life span of a human diploid cell line in culture. Cell Struct. Funct. *2:* 219–227 (1977).

Saksela, E.; Moorhead, P.S.: Aneuploidy in the degenerative phase of serial cultivation of human cell studies. Proc. natn. Acad. Sci. USA *50:* 390–395 (1963).

Salk, D.; Au, K.; Hoehn, H.; Martin, G.M.: Cytogenetics of Werner's syndrome cultured skin: variegated translocation mosaicism. Cytogenet. Cell Genet. *30:* 92–107 (1981a).

Salk, D.; Bryant, E.; Au, K.; Hoehn, H.; Martin, J.G.: Systematic growth studies cocultivation and cell hybridization studies of Werner syndrome culture skin fibroblasts. Hum. Genet. *58:* 310–317 (1981b).

Sambuy, Y.; Bittles, A.H.: The effects of in vitro aging on the composition of the intracellular free amino acid pool of human diploid fibroblasts. Mech. Age. Dev. *20:* 279–287 (1982).

Sambuy, Y.; Bittles, A.H.: The effect of in vitro ageing on the exopeptidases of human diploid fibroblasts. Mech. Age. Dev. *26:* 13–22 (1984).

Sanwal, M.; Muel, A.S.; Chandun, E.; Courtois, Y.; Counis, M.F.: Chromatin condensation and terminal differentiation process in embryonic chicken lens in vivo and in vitro. Expl Cell Res. *167:* 429–439 (1986).

Sarber, R.; Hull, B.; Meriell, C.; Soranno, T.; Bell, E.: Regulation of proliferation of fibroblasts of low and high population doubling levels grown in collagen. Mech. Age. Dev. *17:* 107–117 (1981).

Sax, H.J.; Passano, K.N.: Spontaneous chromosome aberrations in human tissue culture cells. Am. Nat. *95:* 97–102 (1961).

Schachtschabel, D.O.; Wever, J.: Age-related decline in the synthesis of glycosaminoglycans by cultured human fibroblasts (WI-38). Mech. Age. Dev. *8:* 257–264 (1978).

Schafer, J.A.; Pandy, M.; Ferguson, R.; Davis, B.R.: Comparative observation of fibroblasts derived from the papillary and reticular dermis of infants and adults: growth kinetics, packing density at confluence and surface morphology. Mech. Age. Dev. *31:* 275–293 (1985).

Schmid, C.W.; Jelinek, W.R.: The Alu family of dispersed repetive sequences. Science *216:* 1065–1070 (1982).

Schneider, E.L.; Fowlkes, B.J.: Measurement of DNA content and cell volume in senescent human fibroblasts utilizing flow multiparameter single cell analysis. Expl Cell Res. *98:* 298–302 (1976).

Schneider, E.L.; Gilman, B.: Sister chromatid exchanges and aging. III. The effect of donor age on mutagen-induced sister chromatid exchange in human diploid fibroblasts. Hum. Genet. *46:* 57–63 (1979).

Schneider, E.L.; Mitsui, Y.: The relationship between in vitro cellular aging and in vivo human age. Proc. natn. Acad. Sci. USA *73:* 3584–3588 (1976).

Schneider, E.L.; Mitsui, J.; Au, K.S.; Shorr, S.S.: Tissue specific differences in cultured human diploid fibroblasts. Expl Cell Res. *108:* 1–6 (1977).

Schneider, E.L.; Mitsui, J.; Tice, R.; Shorr, S.S.; Braunschweiger, K.: Alteration in cellular RNAs during the in vitro lifespan of cultured human diploid fibroblasts. II. Synthesis and processing of RNA. Mech. Age. Dev. *4:* 449–458 (1975).

Schneider, E.L.; Braunschweiger, K.; Mitsui, Y.: The effect of serum batch on the in vitro lifespans of cell cultures derived from old and young human donors. Expl Cell Res. *115:* 47–52 (1978).

Schofield, R.; Dexter, T.M.; Lord, B.J.; Testz, N.G.: Comparison of haemopoiesis in young and old mice. Mech. Age. Dev. *34:* 1–12 (1986).

Schor, S.L.; Schor, A.M.; Durning, P.; Rushton, G.: Skin fibroblasts obtained from cancer patients display foetal-like migratory behaviour on collagen cells. J. Cell Sci. *73:* 235–244 (1985).

Schroeder, F.; Goetz, J.; Roberts, E.: Age-related alterations in cultured human fibroblasts membrane structure and function. Mech. Age. Dev. *25:* 365–389 (1984).

Schwartz, A.G.: Correlation between species life span and capacity to activate 7,12-dimethylbenzanthracene to a form mutagenic to mammalian cell. Expl Cell Res. *94:* 445–447 (1975).

Schwartz, A.G.; Moore, C.J.: Inverse correlation between species life span and capacity of cultured fibroblasts to bind 7,12-dimethylbenzanthracene to DNA. Expl Cell Res. *109:* 448–450 (1977).

Segal, D.J.; McCoy, E.E.: Studies on Down's syndrome in tissue culture. I. Growth rates and protein contents of fibroblast cultures. J. cell. Physiol. *83:* 85–90 (1974).

Sénéchal, H.; Wahrmann, J.P.; Delain, D.; Macieira-Coelho, A.: Modulation of differentiation in vitro. II. Influence of cell spreading and surface events on myogenesis. In Vitro *20:* 692–698 (1984).

Shakespeare, V.; Buchanan, J.H.: Increased degradation rates of protein in aging human fibroblasts and in cells treated with amino acid analogs. Expl Cell Res. *100:* 1–8 (1976).

Shakespeare, V.; Buchanan, J.H.: Studies on phosphoglucose isomerase from cultured human fibroblasts. Absence of detectable ageing effects on the enzyme. J. cell. Physiol. *94:* 105–114 (1978).

Shakespeare, V.A.; Buchanan, J.H.: Evidence for increased proteolytic activity in aging human fibroblasts. Gerontology *25:* 305–313 (1979).

Shapiro, B.; Lam, L.F.; Fast, L.: Premature senescence in cultured skin fibroblasts from subjects with cystic fibrosis. Science *203:* 1251–1253 (1979).

Shevitz, J.; Jenkins, C.S.P.; Hatcher, V.B.: Fibronectin synthesis and degradation in human fibroblasts with aging. Mech. Age. Dev. *35:* 221-232 (1986).

Shields, R.: Transition probability and the division pattern of WI-38 cells. Cell Biol. int. Rep. *3:* 659–662 (1979).

Shmookler-Reis, R.J.; Goldstein, S.: Loss of reiterated DNA sequences during serial passage of human diploid fibroblasts. Cell *21:* 739–750 (1980).

Shmookler-Reis, R.J.; Goldstein, S.: Variability of DNA methylation patterns during serial passage of human diploid fibroblasts. Proc. natn. Acad. Sci. USA *79:* 3949–3953 (1982a).

Shmookler-Reis, R.J.; Goldstein, S.: Interclonal variation in methylation patterns for expressed and non expressed genes. Nucl. Acids Res. *10:* 4293–4304 (1982b).

Shmookler-Reis, R.J.; Goldstein, S.: Mitochondrial DNA in mortal and immortal human cells. Genome number, integrity and methylation. J. biol. Chem. *258:* 9078–9085 (1983).

Silberberg, R.; Silberberg, M.; Feir, D.: Life cycle of articular cartilage cells: an electron microscopic study of the hip joint of the mouse. Am. J. Anat. *114:* 17–22 (1964).
Siminovitch, L.; Till, J.E.; McCulloch, E.A.: Decline in colony-forming ability of marrow cells subjected to serial transplantation into irradiated mice. J. cell. comp. Physiol. *64:* 23–29 (1964).
Simnett, J.D.; Morley, A.R.: Factors controlling growth of prostatic epithelium. A comparison of mitotic activity in mice of different ages in vitro and in organ culture. Expl Cell Res. *46:* 29–36 (1967).
Simon, M.; Green, H.: Participation of membrane-associated proteins in the formation of the cross-linked envelope of the keratinocyte. Cell *36:* 827–834 (1984).
Simons, J.W.I.M.: The use of frequency distributions of cell diameters to characterize cell populations in tissue culture. Expl Cell Res. *45:* 336–350 (1967).
Simons, J.W.; Van den Broek, C.: Comparison of ageing in vitro and ageing in vivo by means of cell size analysis using a Coulter counter. Gerontologia *16:* 340–351 (1970).
Sisken, J.E.; Bonner, S.V.: On the duration of mitotic stages in senescing human fibroblasts in culture. Mech. Age. Dev. *11:* 191–197 (1979).
Skalli, O.; Ropraz, P.; Trzeciak, A.; Beuzonans, G.; Gillesseu, D.; Gabbiani, G.: A monoclonal antibody against alpha-smooth muscle actin. A new probe for smooth muscle differentiation. J. Cell Biol. *103:* 2787–2796 (1986).
Skehan, P.; Friedman, S.J.: Cellular senescence and the analysis of generation time distribution. Cell Biol. int. Rep. *3:* 535–542 (1979).
Slayback, J.R.B.; Cheung, L.W.Y.; Geger, R.P.: Comparative effects of human platelet growth factor on the growth and morphology of human fetal and adult diploid fibroblasts. Expl Cell Res. *110:* 462–466 (1977).
Sluke, G.; Schachtschabel, D.O.; Wever, J.: Age-related changes in the distribution pattern of glycosaminoglycans synthesized by cultured human diploid fibroblasts (WI-38). Mech. Age. Dev. *16:* 19–27 (1981).
Smith, B.T.; Torday, J.S.; Giroud, C.H.P.: The growth promoting effect of cortisol on human fetal lung cells. Steroids *22:* 515–524 (1973).
Smith, J.A.; Martin, L.: Do cell cycle? Proc. natn. Acad. Sci. USA *70:* 1263–1267 (1973).
Smith, J.R.: Application of the theory of transition probability in 'ageing' WI-38 cells; similar behaviour of clonogenic cells from early and late passage cultures. Cell. Biol. int. Rep. *1:* 283–289 (1977).
Smith, J.R.; Braunschweiger, K.I.; Growth of human embryonic fibroblasts at clonal density. Concordance with results from mass cultures. J. cell. Physiol. *98:* 597–604 (1979).
Smith, J.R.; Pereira-Smith, O.M.; Schneider, E.L.: Colony size distributions as a mesure of in vivo and in vitro aging. Proc. natn. Acad. Sci. USA *75:* 1353–1356 (1978).
Smith, J.R.; Whitney, R.G.: Intraclonal variation in proliferation potential of human diploid fibroblasts. Stochastic mechanism for cellular aging. Science *207:* 82–84 (1979).
Smith, P.J.; Paterson, M.C.: Enhanced radiosensitivity and defective DNA repair in cultured fibroblasts derived from Rothmund Thomson syndrome patients. Mutation Res. *94:* 213–228 (1982).
Someville, M.; Houben, A.; Raes, M.; Houbion, A.; Henin, V.; Remacle, J.: Alteration of enzymes in aging human fibroblasts in culture. III. Modifications of superoxide dismutase as an environmental and reversible process. Mech. Age. Dev. *29:* 35–51 (1985).

Sorrentino, J.A.; Mills, A.J.T.: Structural comparisons of fibronectins isolated from early and late passage cells. Mech. Age. Dev. *28:* 83–97 (1984).

Soukupova, M.; Holeckova, E.: The latent period of explanted organs of newborn adult and senile rats. Expl Cell Res. *33:* 361–367 (1964).

Spataro, A.C.; Bosmann, H.B.; Myers-Robfogel, M.W.: Sialyltransferase activities of aging diploid fibroblasts. Biochim. biophys. Acta *553:* 378–387 (1979).

Srivastava, B.J.S.: Changes in enzyme activity during cultivation of human cells in vitro. Expl Cell Res. *80:* 305–312 (1973).

Srivastava, A.; Norris, J.S.; Shmookler-Reis, R.J.; Goldstein, S.: c-Ha-ras-1 proto-oncogene amplification and overexpression during the limited replicative life span of normal human fibroblasts. J. biol. Chem. *260:* 6404–6409 (1985).

Staiano-Coico, L.; Darzynkiewicz, Z.; Melamed, M.R.; Weksler, M.: Changes in DNA content of human blood mononuclear cells with senescence. Cytometry *3:* 79–83 (1982).

Stanley, J.F.; Pye, D.; MacGregor, A.: Comparison of doubling numbers attained by cultured animal cells with life span of species. Nature, Lond. *255:* 158–159 (1975).

Stanners, C.P.; Till, J.E.: DNA synthesis in individual L strain mouse cells. Biochim. biophys. Acta *37:* 406–419 (1960).

Stanulis-Praeger, B.M.: Filopodia number increases with age and quiescence in populations of normal WI-38 cells and is correlated with drug-induced changes in proliferation on both normal and transformed populations. Mech. Age. Dev. *33:* 221–235 (1986).

Stefanini, M.; Orecchia, G.; Rabbiosi, G.; Nuzzo, F.: Altered cellular response to UV irradiation in a patient affected by premature aging. Hum. Genet. *73:* 189–192 (1986).

Stehelin, D.; Guntaka, R.V.; Varmus, H.E.; Bishop, J.M.: Purification of DNA complementary to nucleotide sequences required for neoplastic transformation of fibroblasts by avian sarcoma viruses. J. mol. Biol. *101:* 349–365 (1976).

Stein, G.H.: DNA-binding proteins in young and senescent normal human fibroblasts. Expl Cell Res. *90:* 237–248 (1975).

Stein, G.H.: Human diploid fibroblasts (HDF) can induce DNA synthesis in cycling HDF but not in quiescent HDF or senescent HDF. Expl Cell Res. *144:* 468–471 (1983).

Stein, G.H.: SV_{40}-transformed human fibroblasts. Evidence for cellular aging in precrisis cells. J. cell. Physiol. *125:* 36–44 (1985).

Stein, G.H.; Yanishevsky, R.M.: Entry into S phase is inhibited in two immortal cell lines fused to senescent human diploid cells. Expl Cell Res. *120:* 155–165 (1979).

Stein, G.H.; Yanishevsky, R.M.. Quiescent human diploid cells can inhibit entry into S phase in replicative nuclei in heterodikaryons. Proc. natn. Acad. Sci. USA *78:* 3025–3029 (1981).

Stein, G.H.; Atkins, L.: Membrane-associated inhibitor of DNA synthesis in senescent human diploid fibroblasts. Characterization and comparison to quiescent cell inhibitor. Proc. natn. Acad. Sci. USA *83:* 9030–9034 (1986).

Stein, G.H.; Yanishevsky, R.M.; Gordon, L.; Beeson, M.: Carcinogen-transformed human cells are inhibited from entry into S phase by fusion to senescent cells but cells transformed by DNA tumor viruses overcome the inhibition. Proc. natn. Acad. Sci. USA *79:* 5287–5292 (1982).

Stein, G.H.; Atkins, L.; Beeson, M.; Gordon, L.: Quiescent human diploid fibroblasts.

Common mechanism for inhibition of DNA replication in density-inhibited and serum-deprived cells. Expl Cell Res. *162:* 255–260 (1986).

Stein, G.S.; Burtner, D.L.: Gene activation in human diploid cells. Age dependent modifications in the stability of messenger RNAs for non-histone chromosome proteins. Biochim. biophys. Acta *390:* 56–68 (1975).

Steinhardt, M.: Effect of donor age on clonal differentiation of human skin fibroblast in vitro. Gerontology *31:* 27–38 (1985).

Stenkvist, B.: Long-term cultivation of human and bovine fibroblastic cells morphologically transformed in vitro by Rous sarcoma virus. Acta pathol. microbiol. scand. *67:* 67–82 (1966).

Strehler, B.L.: Roles and mechanisms of rDNA changes during aging; in Bergsma, Harrison, Genetic effects of aging. The National foundation – March of Dimes Birth defects, pp. 445–462 (Liss, New York 1978).

Strehler, B.; Chang, M.P.: Loss of hybridizable ribosomal DNA from human postmitotic tissues during aging. II. Age dependent loss in human cerebral cortex – hippocampal and somato sensory cortex comparison. Mech. Age. Dev. *11:* 375–382 (1979).

Strehler, B.L.; Chang, M.P.; Johnson, L.K.: Loss of hybridizable ribosomal DNA from human postmitotic tissues during aging. I. Age-dependent loss in human myocardium. Mech. Age. Dev. *11:* 371–378 (1977).

Sugarman, B.; Munro, H.N.: Altered accumulation of zinc by aging human fibroblasts in culture. Life Sci. *26:* 915–920 (1980).

Sun, A.S.; Aggarwal, B.B.; Packer, L.: Enzyme levels of normal human cells: aging in culture. Archs Biochem. Biophys. *170:* 1–11 (1975).

Sun, A.S.; Alvarez, L.J.; Reinach, P.S.; Rubin, E.: 5′-Nucleotidase levels in normal and virus transformed cells. Lab. Invest. *41:* 1–4 (1979).

Suzuki, F.; Watenabe, F.; Horikawa, U.: Repair of X-ray induced DNA damage in aging human diploid cells. Expl Cell Res. *127:* 299–307 (1980).

Swirenga, S.H.H.; Whitfield, J.F.; Gillan, D.J.: Alteration by malignant transformation of the calcium requirements for cell proliferation in vitro. J. natn. Cancer Inst. *57:* 125–129 (1976).

Swierenga, S.H.H.; Whitfield, J.F.; Boynton, L.: Age related and carcinogen-induced alterations of the extracellular growth factor requirements for cell proliferation in vitro. J. cell. Physiol. *94:* 171–180 (1978).

Swim, H.E.; Parker, R.F.: Culture characteristics of human fibroblasts propagated serially. Am. J. Hyg. *66:* 235–243 (1957).

Takahashi, S.; Beifter, S.; Davidson, A.: Enzymes of the alphaglutamyl cycle in 'aging' WI-38 fibroblasts and in Hela S_3 cells. Biochim. biophys. Acta *522:* 63–73 (1978).

Takahashi, S.; Nakagawa, S.; Zeydel, M.; Bhargava, G.: Ammonia production by IMR-90 fibroblast cultures: effects of ammonia on glutathiase gamma-glutamyl transpeptidase, lysosomal enzymes and cell division. J. cell. Physiol.*125:* 107–114 (1985).

Takahashi, S.; Zeydel, M.: Alpha-glutamyl transpeptidase and glutathione in aging IMR-90 fibroblasts and in differentiating 3T3 L1 preadipocytes. Archs Biochem. Biophys. *214:* 260–268 (1982).

Takeuchi, F.; Goto, M.; Mijamoto, T.: Normal thermostability of hypoxanthine-guanine phosphoribosyltransferase in eryhtrocytes from Werner' syndrome patients. Gerontology *32:* 196–201 (1986).

Takeuchi, F.; Hanaoka, F.; Goto, M.; Akaoka, J.; Hori, T.; Yamada, M.; Miyamoto, T.: Altered frequency of initiation sites of DNA replication in Werner's syndrome cells. Hum. Genet. *60:* 365–369 (1982).

Tan, J.H.; Chou, E.L.; Lundh, L.N.: Regulation of chromosome 21-directed antiviral gene(s) as a consequence of age. Nature, Lond. *257:* 310–311 (1975).

Tappel, A.L.: Lipid peroxidation damage to cell components. Fed. Proc. *32:* 1870–1874 (1973).

Tas, S.; Tam, C.F.; Walford, R.L.: Disulfide bonds and the structure of the chromatin complex in relation to aging. Mech. Age. Dev. *12:* 65–80 (1980).

Tassin, J.; Malaise, E.; Courtois, Y.: Human lens cells have an in vitro proliferative capacity inversely proportional to the donor age. Expl Cell Res. *123:* 388–392 (1979).

Taylor, J.H.: Sister chromatid exchanges in tritium-labelled chromosomes. Genetics *43:* 515–529 (1958).

Terzi, M.; Hawkins, T.S.C.: Chromosomal variation and the establishment of somatic cell lines in vitro. Nature, Lond. *253:* 361–362 (1975).

Thaw, H.H.; Collins, V.P.; Brunk, U.T.: Influence of oxigen tension pro-oxidants and antioxidants on the formation of lipid peroxidation products (lipofuscin) in individual cultivated human glial cells. Mech. Age. Dev. *24:* 211–223 (1984).

Thurston, W.P.; Weeks, J.R.: The mathematics of three-dimensional manifolds. Scient. Am. *July:* 94–106 (1984).

Tjio, J.H.; Nichols, W.: History and present status of human chromosome studies. In Vitro Cell. dev. Biol. *21:* 305–313 (1985).

Todaro, G.J.; Green, H.: Quantitative studies of the growth of mouse embryo cells in culture and their development into established lines. J. Cell Biol. *17:* 299–313 (1963).

Tollefsbol, T.; Zaum, M.R.; Gracy, R.: Increased lability of triosephosphate isomerase in progeria and Werner's syndrome fibroblast. Mech. Age. Dev. *20:* 93–101 (1982).

Tomkins, G.A.; Stanbridge, E.J.; Hayflick, L.: Viral probes of ageing in the human diploid cell strain WI-38. Proc. Soc. exp. Biol. Med. *146:* 385–390 (1974).

Torrelio, B.M.; Paz, M.A.; Gallop, P.M.: Modulation of purine synthesis and phosphoribosyl pyrophosphate content in human fibroblasts at different population doublings. Mech. Age. Dev. *19:* 147–158 (1982).

Treton, J.A.; Courtois, Y.: Evolution of the distribution, proliferation and ultraviolet repair capacity of rat lens epithelial cells as a function of maturation and aging. Mech. Age. Dev. *15:* 251–267 (1981).

Tsuji, Y.; Ide, T.; Ishibashi, S.; Nishikawa, K.: Loss of responsiveness in senescent human TIG-1-cells to the DNA synthesis-inducing effect of various growth factors. Mech. Age. Dev. *27:* 219–232 (1984).

Tsuji, Y.; Ishibashi, S.; Ide, T.: Induction of heat shock proteins in young and senescent human diploid fibroblasts. Mech. Age. Dev. *36:* 155–160 (1986).

Udupa, K.B.; Lipschitz, D.A.: Erythropoiesis in the aged mouse. I. Response to stimulation in vivo. J. Lab. clin. Med. *103:* 574–580 (1984).

Van't Hof, J.: PEA (pisum sativum) cells arrested in G2 have nascent DNA with breaks between replicons and replication clusters. Expl Cell Res. *129:* 231–237 (1980).

Van't Hof, J.; Bjerknes, C.A.: Cells of Pea (pisum sativum) that differentiate from G2 phase have extrachromosomal DNA. Mol. cell. Biol. *2:* 339–345 (1982).

Viceps-Madore, D.; Cristofalo, V.J.: Age-associated changes in glutamine synthetase activity in WI-38 cells. Mech. Age. Dev. *8:* 43–50 (1978).
Vijg, J.; Mullaart, E.P.; Schans, G. van der; Lohman, P.H.M.; Knook, D.L.: Kinetics of ultraviolet induced DNA excision repair in rat and human fibroblasts. Mutation Res. *132:* 129–238 (1984).
Vijg, J.; Mullaart, E.; Berends, F.; Lohman, P.H.M.; Knook, D.L.: UV-induced DNA excision repair in rat fibroblasts during immortalization and terminal differentiation in vitro. Expl Cell Res. *167:* 517–530 (1986).
Vincent, R.A., Jr.; Huang, P.C.: The proportion of cells labeled with tritiated thymidine as a function of population doubling level in cultures of fetal, adult, mutant and tumor origin. Exp. Cell Res. *102:* 31–42 (1976).
Vogel, K.G.; Kelley, R.O.; Steward, C.: Loss of organized fibronectin matrix from the surface of aging diploid fibroblasts. Mech. Age. Dev. *16:* 295–303 (1981a).
Vogel, K.G.; Kendall, V.F.; Spanien, R.E.: Glycosaminoglycan synthesis and composition in human fibroblasts during in vitro cellular aging (IMR-90). J. cell. Physiol. *107:* 271–281 (1981b).
Vogel, K.G.; Pitcher, D.E.: Sulfated proteoglycans produced by human lung fibroblasts during in vitro cellular aging. Eur. J. Cell Biol. *29:* 61–67 (1982).
Volicer, L.; West, C.D.; Chase, A.R.; Greene, L.: Beta-adrenergic receptor sensitivity in cultured vascular smooth muscle cells. Effect of age and a dietary restriction. Mech. Age. Dev. *21:* 283–293 (1983).
Vos, O.; Dolmans, M.J.A.S.: Self-renewal of colony forming units (CFU) in serial bone marrow transplantation experiments. Cell Tiss. Kinet. *5:* 371–385 (1972).
Vracko, R.; McFarland, B.H.: Lifespan of diabetic and non-diabetic fibroblasts in vitro. Results of replicate determinations. Expl Cell Res. *129:* 345–350 (1980).
Vracko, R.; McFarland, B.H.; Pecoraro, R.E.: Seeding efficiency, plating efficiency and population doublings of human skin fibroblast-like cells. Results of replicate testing. In Vitro *19:* 504–514 (1983).
Walford, R.: Multigene families, hystocompatibility systems, transformation, meiosis, stem cells and DNA repair. Mech. Age. Dev. *9:* 19–26 (1979).
Walford, R.L.; Jawaid, S.Q.; Naeim, F.: Evidence for in vitro senescence of T-lymphocytes cultured from normal human peripheral blood. Age *4:* 67–70 (1981).
Wang, E.: A 57000 mol-wt protein uniquely present in nonproliferating cells and senescent human fibroblasts. J. Cell Biol. *100:* 545–551 (1985a).
Wang, E.: Rapid disappearance of statin a nonproliferating and senescent cell-specific protein upon reentring the process of cell cycling. J. Cell Biol. *101:* 1695–1701 (1985b).
Wang, E.: Are cross-bridging structures involved in the bundle formation of intermediate filaments and the decrease in locomotion that accompany cell aging? J. Cell Biol. *100:* 1466–1473 (1985c).
Wang, K.M.; Rose, N.R.; Bartholomew, E.A.; Balzer, M.; Berde, K.; Foldvary, M.: Changes of enzymic activities in human diploid cell line WI-38 at various passages. Expl Cell Res. *61:* 357–364 (1970).
Wahrmann, J.P.; Delain, D.; Bournoutian, C.; Macieira-Coelho, A.: Modulation of differentiation in vitro. I. Influence of the attachment surface on myogenesis. In Vitro *17:* 752–762 (1981).
Wasteson, Å.; Uthne, K.; Westermark, B.: A novel assay for the biosynthesis of sulphated

polysaccharide and its application to studies on the effects of somatomedin on cultured cells. Biochem. J. *136:* 1069–1074 (1973).

Waters, H.; Walford, R.L.: Latent periods for outgrowth of human skin explants as a function of age. J. Geront. *25:* 381–383 (1970).

Watt, F.M.; Green, H.: Involucrin synthesis is correlated with cell size in human epidermal cultures. J. Cell Biol. *90:* 738–742 (1981).

Weiss, C.; Ultrastructural characteristics of osteoarthritis. Fed. Proc. *32:* 1459–1466 (1975).

Weisman, A.: Essays upon heredity and kindred biological problems, vol. I (Clarendon Press, Oxford 1891).

Westermark, B.; Blomqvist, E.: Stimulation of fibroblast migration by epidermal growth factor. Cell Biol. int. Rep. *4:* 649–654 (1980).

Westermark, B.; Wasteson, Å.: The response of cultured human normal glial cells to growth factors; in Advances in metabolic disorders, vol. 8, pp. 85–100 (Academic Press, New York 1975).

Wever, J.; Schachtschabel, D.O.; Sluke, G.; Wever, G.: Effect of short or long term treatment with exogenous glycosaminoglycans on growth and glycosaminoglycan synthesis of human fibroblasts (WI-38) in culture. Mech. Age. Dev. *14:* 89–99 (1980).

Wharton, W.: Newborn human skin fibroblasts senesce in vitro without acquiring adult growth factor requirements. Expl Cell Res. *154:* 310–314 (1984).

Wharton, S.A.; Riley, P.A.: The suppression of endogenous protein degradation by fractions of foetal calf serum. Dialysed serum is less able to suppress degradation in aged cells. Cell Biochem. Funct. *4:* 189–196 (1986).

Whatley, S.A.; Hill, B.T.: The relationship between DNA content, cell volume and growth potential in ageing human embryonic mesenchymal cells. Cell Biol. int. Rep. *3:* 671–683 (1979).

White, F.A.; Bunn, C.L.: Restriction enzyme analysis of mitochondrial DNA in aging human cells. Mech. Age. Dev. *30:* 153–168 (1985).

Wieser, R.J.; Oesch, F.: Contact inhibition of growth of human diploid fibroblasts by immobilized plasma membrane glycoproteins. J. Cell Biol. *103:* 361–367 (1986).

Williams, J.R.: Role of DNA repair in cell inactivation, aging and transformation. A selective review, a speculative model. Adv. radiat. Biol. *6:* 161–210 (1976).

Williamson, A.R.; Askonas, B.A.: Senescence of an antibody-forming cell clone. Nature, Lond. *238:* 337–339 (1972).

Wojtyk, R.J.; Goldstein, S.: Fidelity of protein synthesis does not decline during aging of cultured human fibroblasts. J. cell. Physiol. *103:* 299–303 (1980).

Wojtyk, R.J.; Goldstein, S.: Clonal selection in cultured human fibroblasts. Role of protein synthetic errors. J. Cell Biol. *95:* 704–710 (1982).

Wolf, S.; Sameshima, M.; Liebhaber, S.A.; Schlessinger, D.: Regulation of ribosomal ribonucleic acid levels in growing, ^{3}H-arrested and crisis-phase WI-38 human diploid fibroblasts. Biochemistry *19:* 3484–3490 (1980).

Woods, C.M.; Lazarides, E.: Spectrin assembly in avian erythroid development is determined by competing reactions of subunit homo- and hetero-oligomerization. Nature, Lond. *321:* 85–89 (1986).

Wright, W.E.; Hayflick, L.: Nuclear control of cellular aging demonstrated by hybridization of annucleate and whole cultured normal human fibroblasts. Expl Cell Res. *96:* 113–121 (1975).

Yagi, T.: DNA repair ability of cultured cells derived from mouse embryos in comparison with human cells. Mutation Res. *96:* 89–98 (1982).

Yamagishi, H.; Kunisada, T.; Iwakura, Y.; Nishimuno, Y.; Ogiso, Y.; Matsuhiro, A.: Emergence of the extrachromosomal circular DNA complexes as one of the earliest signals of cellular differentiation in the early development of mouse embryo. Develop. Growth Different. *25:* 563–569 (1983).

Yamanak, N.; Deamer, D.: Superoxide dismutase activity in WI-38 cell cultures. Effects of age, trypsinization and SV_{40} transformation. Physiol. Chem. Phys. *6:* 95–106 (1974).

Yamamoto, K.; Yamamoto, M.; Ooka, H.: Cell surface changes associated with aging of chick embryo fibroblasts in culture. Expl Cell Res. *108:* 87–93 (1977).

Yanishevsky, R.; Carrano, A.V.: Prematurely condensed chromosomes of dividing and non dividing cells in aging human cell cultures. Expl Cell Res. *90:* 169–174 (1975).

Yao, M.C.; Choi, J.; Yokoyama, S.; Austerberry, C.F.; Yat, C.H.: DNA elimination in tetrahymena. A developmental process involving extensive breakage and rejoining of DNA at defined sites. Cell *36:* 433–440 (1984).

Yaron, M.; Yaron, I.; Caspi, D.; Smetan, O.; Eylan, E.; Zor, V.: RNA and DNA viral stimulation of prostaglandin E production by human synovial fibroblasts. Arthritis Rheum. *24:* 1582–1584 (1981).

Yatscoff, R.W.; Goldstein, S.; Freedman, K.B.: Conservation of genes coding for protein synthesized in human mitochondria. Som. Cell Genet. *4:* 633–645 (1978).

Yuan, P.M.; Talent, J.M.; Gracy, R.W.: Molecular basis for the accumulation of acid isozymes of triosephosphate isomerase in aging. Mech. Age. Dev. *17:* 151–162 (1981).

Zavala, C.; Herner, G.; Fialkow, P.J.: Evidence for selection in cultured diploid fibroblasts strains. Expl Cell Res. *117:* 137–144 (1978).

Zorzoli, A.: The influence of age on phosphatase activity in the mouse. J. Geront. *10:* 156–164 (1955).

Subject Index